Introduction to the Techniques of Chemistry

"What you have been obliged to discover for yourself leaves a path in your mind which you can use again when the need arises."

--Lichtenberg, 18th-century German physicist

HAMPDEN-SYDNEY COLLEGE

W. W. Porterfield

2015

CHEM 151 LAB ATTENDANCE SHEET

Fall 2007

NAME Irv Gratch

DESK NO. 313

Fill in spaces below with time in/time out; for example: 1:30 / 4:45

	Mon	Tue	Wed	Thu	Fri
A29	/	/	1:30 / 4:30	/	/
S5	/	/	1:30 / 4:45	2:00 / 3:00	/
S12	/	/	1:30 / 5:00	/	/
S19	/	/	1:30 / 4:15	/	/
S26	10:00 / 10:15	/	1:30 / 4:40	2:15 / 2:45	/
O3	/	/	1:30 / 4:30	/	/
O10	/	/	1:30 / 4:15	/	/
O17	/	/	1:30 / 5:00	/	/
O24	/	/	1:30 / 3:10	1:45 / 3:00	/
O31	/	/	1:30 / 4:15	/	/
N7	/	/	1:30 / 3:00	/	/
N14	/	/	1:30 / 4:45	/	/
N28	/	/	1:30 / 3:15	/	/

TABLE OF CONTENTS

Schedule 5

General Instructions and Safety Precautions 7

Weight and Balance Instructions 10

Volume and Glassware Instructions 14

Solution Concentrations and Preparation 17

Experimental Errors and Precision 20

Laboratory Notebook Instructions 30

Check-in Instructions and Apparatus Diagrams 34

Fall Semester Project: Coordination Compound Synthesis and Analysis 38

Part I: Synthesis 38

Translating Literature Procedures 39

Journal-Type Procedures 40

Inorganic Syntheses Procedures 40

Safety Precautions 41

Common Techniques 41

Writing Up a Synthesis Procedure 42

Calculations and Project Summary 46

Part II: Analysis 47

Analytical Measurement Accuracy 47

Volumetric Metal Analysis 48

EDTA Standardization 52

Metal-EDTA Titrations 54

Gravimetric Halide Analysis 55

Atomic Emission Metal Analysis 59

Final Summary 63

Ending the Semester 63

Spring Semester Project: Identification of Organic Compounds 65
Organic Liquid Mixture 65
Boiling Point and Distillation 65
Gas Chromatography and Purity 69
Sieving the Table for Molecular Structures 71
Spectroscopic Properties Introduction 71
Naming Organic Compounds 74
^{13}C Nuclear Magnetic Resonance 76
How the NMR Instrument Gives Us Spectra 82
The FT-NMR Technique Quiz 86
Sample Preparation 87
Interpreting the spectrum and writing it up 88
Infrared Spectroscopy 92
Fourier-Transform Infrared Instrumentation 97
The FT-IR Technique Quiz 101
Running the IR Spectrum 102
Notebook Writeup 103
^{1}H Nuclear Magnetic Resonance 104
The ^{1}H NMR Quiz 112
Figuring out Structures with Benzene Rings from ^{1}H NMR Spectra 113
Notebook Writeup 116
Summary for Liquid Identification 117
Table of Possible Organic Liquids 118
Identification of Solid Organic Acids 127
Melting Point 127
Recrystallization 130
Neutralization Equivalent and pK_a 131
^{1}H NMR for Your Solid Acid 136
IR for Your Solid Acid 136
Table of Possible Organic Acids 137
Summary for Solid Acid 138
Final Summary Writeup 138
Laboratory Checkout Instructions 140

SCHEDULE

Fall Semester Project: Synthesis and Quantitative Analysis Techniques—Preparation and Analysis of a Metal Coordination Compound

You will receive the name and formula of a compound you will be expected to make; the compound will be a combination of a metal halide and another compound (usually organic) called a ligand. You will use the library to find a method of preparing the compound and have the method approved by a faculty member. You will then prepare 10 grams of the compound for analysis and store it, bottled, in a desiccator for analysis.

In the analysis of the compound for its elemental composition, you will obtain weight % halide by a gravimetric analysis, weight % metal by a compleximetric titration, and demonstrate your mastery of the techniques by analyzing a known compound for practice before attacking your own coordination compound. You will also determine the weight % metal by an instrumental technique, atomic emission spectroscopy.

Note in the lab lecture schedule below that your notebook will be due twice, once at the end of the synthesis and again at the end of the analysis. In addition, to give you an idea of the pace at which various lab activities should be accomplished, 3x5 cards will be due occasionally for intermediate results. The results themselves won't be graded till we look at the notebook at the end of the semester, but we will charge a point or two for late submission of the card.

Week	Date	Topic for lecture	Notebook or 3x5 card due
1	____	checkin and project outline	
2	____	chemical literature use and synthetic techniques	
3	____	synthetic and storage techniques; yield calculations	
4	____	volumetric techniques and glassware use	NOTEBOOK DUE
5	____	EDTA solution preparation and standardization	
6	____	metal-EDTA titrations	EDTA standardization
7	____	-- [no lecture]	practice M-EDTA
8	____	gravimetric halide analysis	coord cpd M-EDTA
9	____	-- [no lecture]	
10	____	atomic emission (AE) spectroscopy	practice gravimetric X
11	____	quantitative elemental analysis calculations	
12	____	-- [no lecture]	coord cpd gravimetric X
13	____	-- [no lecture]	
14	____	-- [no lecture]	NOTEBOOK DUE FRIDAY

Your breakage deposit is refundable (except for the cost of this manual and a lab notebook), but *only* if you check out at the end of lab *on time*. We won't grade lab notebooks unless they contain a checkout slip or a preregistration slip for Chem 152 from the stockroom manager! The deadline for checking out if you're leaving the lab, obtaining that slip, and turning in your notebook is 4:00 PM on Friday of week 14—that is, at least one day after the last day for work in the lab.

Spring Semester Project: Identification of Unknown Organic Compounds

You will receive a bottle containing about 40 mL of a mixture of two liquid organic compounds and another bottle containing an unknown solid organic acid. You will separate the liquids by fractional distillation and obtain the boiling point range of each in the process. You will use gas chromatography to check each distillation fraction for purity. A table of candidate compounds is provided, which you can sieve to get an initial list of candidate compounds for each distillation fraction.

With a list of several candidate compounds for each fraction, you will use introductory organic nomenclature rules to work out molecular structures and chemical bonding patterns for each candidate. On the basis of these structures you will predict the ^{13}C nuclear magnetic resonance (NMR) spectrum for each candidate structure, have the spectrum run, and use the experimental spectrum to refine your list of candidate structures by establishing the symmetry of the carbon-atom framework. You will predict the infrared (IR) spectrum for each remaining candidate structure, run that spectrum yourself, and use the experimental spectrum to further refine your list of candidates by establishing the nature of functional groups in the molecule, including a computer database search of possible spectra. You will predict the ^{1}H NMR spectrum for each remaining candidate, have the spectrum run, and use the experimental spectrum to further refine your list of candidates by establishing the functional-group environment, number, and neighbor-H environment of each type of hydrogen atom present—which should establish the identity of each distillation-fraction molecule. If you finish early, some interesting advanced instrument techniques can be used for extra credit.

For the unknown acid, you will obtain a melting point and use that value as a sieve for a list of possible acids, whose properties you will obtain from handbooks. You will narrow the list by obtaining a neutralization equivalent for the acid by titration, and finally determine its identity from its ^{1}H NMR spectrum and optionally from its IR spectrum.

Week	Date	Topic for lecture	Due dates
1	____	liq: distillation/sol: MP, recrystallization	
2	____	liq: gas chromatography/sol: neutralization equiv.	
3	____	organic structures and nomenclature	3x5 card due with BPs
4	____	basic NMR spectroscopy for ^{13}C	
5	____	NMR instrumentation	
6	____	basic IR spectroscopy	
7	____	IR instrumentation	3x5 card due with ^{13}C data
8	____	basic NMR spectroscopy for ^{1}H	
9	____	making up and standardizing NaOH soln	3x5 card due with IR data
10	____	-- [no lecture]	
11	____	^{1}H NMR spectroscopy for benzene rings	3x5 card due with liquid unkns
12	____	-- [no lecture]	
13	____	-- [no lecture]	
14	____	-- [no lecture]	NOTEBOOK DUE FRIDAY

Your breakage deposit is refundable (except for the cost of this manual and two lab notebooks), but *only* if you check out at the end of lab *on time*. We won't grade lab notebooks unless they contain a checkout slip from the stockroom manager! The deadline for checking out is your last working day in lab, but the notebook deadline is 4:00 PM on Friday of the last week of lab.

GENERAL INSTRUCTIONS AND SAFETY PRECAUTIONS

LABORATORY HOURS: 9:00 AM to 5:00 PM, Monday through Friday.

The laboratory will not be open and *work is not permitted* outside these stipulated periods. **Notebooks are due at 5:00 PM of the deadline lab day for synthesis, or 4:00 PM on Friday of the last week**, and will be graded down sharply for late submission, even if it's only a few minutes! Notebooks late by less than 24 hours will be charged 30 points; those late by more than 24 hours will not be accepted.

LABORATORY ATTENDANCE: Attendance is required on your scheduled day. There will be a short lab lecture at the beginning of each lab period (with only a few exceptions) starting at 1:30. *You are responsible for recording your lab attendance times* (including the lab lecture) in an attendance log notebook kept in the laboratory, which has a page with your name. A typical attendance-log page was shown on page 2; note that time spent in the library during the first week *counts as lab time and should be shown*. There is no formal grade deduction for absences, but long experience makes it absolutely certain that if you miss more than one or two of these two-per-week attendance checks you will fall behind in work on your lab project and your grade will suffer. We'll use the attendance checks to alert us to who is falling behind.

LAB LECTURES: In the fall there will be about 30 minutes worth of lab lecture each day (a bit longer in the spring); the lectures are intended to help you organize your effort. As noted above, you must attend, *and you must take notes*. Take them in the back of your notebook, and print a title for each lecture so that we can check them when you turn in the notebook at the end of the project. ***There should, in general, be about a page of notes for each lecture.*** We will deduct 3 points (from a weekly project credit of 20 points) for each missing set of notes for a single lecture. Start each lecture's notes on a new page. The sequence of lectures for the two semesters' projects is given on the two previous pages.

LABORATORY PREPARATION AND ABSENCES: These projects are designed so that they can be completed in one laboratory afternoon per week if (but only if) each afternoon's work is thoroughly and carefully planned before you report to the laboratory. ***Procedures and data pages should be written up in the notebook before reporting to the laboratory.*** Lab time must be spent in conducting experiments, not in deciding what to do or in writing it out. If you fail to organize your time appropriately, the result will be greatly increased time spent in the lab; planning is important! If you should fall behind, you're permitted to work in the lab in the morning or on other afternoons, but only on specific experiments that you have had a lab lecture on and written up in your notebook.

The attendance log and your responsibility for your record in it (including any extra time) have been noted above. One problem that arises is athletic participation; obviously, some of you will need to leave about 3:30 for practice. This is OK, but do two things: first, record the time you *do* spend in the attendance log; second, work out a plan for making up the lab time you miss, either in the morning or another afternoon. It will be necessary to plan your work particularly carefully. Naturally, consideration will be given to extensions of deadlines when athletics or other activities require your absence from the campus — but such extensions can't be guaranteed for every case, and although deadlines can be extended, remember that semesters can't.

PLEDGED WORK: Hampden-Sydney has a time-honored and effective honor system. The work of this laboratory must be appropriately pledged. You may receive help only from authorized sources; these include the chemistry faculty, student lab assistants, other students taking the lab at the same time, and lab manuals or reference books. When reference sources (including people) are used *in any manner,* full reference to the name of the source (page no., etc.) and a summary of all information obtained must appear in your laboratory notebook. *Unauthorized* sources include ALL students not referred to above and any notes, lab notebooks, or other materials belonging to or obtained from other persons. At the end of each semester project you must write out and sign the following pledge:

I pledge on my honor that

1) I have not received any unauthorized help on this project.

2) I have recorded all experiments that I performed in my notebook without duplication or recopying; all entries were made immediately as work was performed.

3) All experiments recorded result from actual experimentation in the laboratory.

4) I have not altered the data in any manner to correspond more closely to expected values.

5) I have properly disposed of all chemical waste generated this semester.

GRADING: At the end of the semester your notebook will be graded on the basis of 20 points per week of work, or 300 points for the project. The lab is a separate course with its own course letter grade, and **this grade will depend entirely on your notebook.** The notebook will be graded for style and to some extent for neatness, but primarily for its content: what specific experiments did you do, how thoroughly and how well did you do them, and how carefully did you think about what the results mean for your overall project?

There are some additional factors that can influence your grade; we've already noted that points will be deducted for missing intermediate deadlines and for failing to keep lab lecture notes. There are two other important possibilities:

The good news: Five points extra credit will be given for each laboratory week ahead of the deadline that you skip by turning your completed project notebook in early. This is an exceptionally good way to remedy minor defects in your grade.

The bad news: Lowering of your laboratory grade on a given experiment will occur if you violate the laboratory rules in the next section. We have attempted to minimize these rules; they are necessary for the smooth operation of the laboratory. We are prepared to be strict in their enforcement.

LABORATORY RULES:

1) **You MUST wear prescription or safety glasses and shoes** (not flip-flops) when in the laboratory, whether you are "doing chemistry" or not—and in the hallway if you're carrying chemicals.

2) **You MUST know and follow sound safety practice**. Some general safety precautions and basic first-aid guidelines are listed in the next section, and the manual's discussion of each project will include some specific safety considerations. You are responsible for all of these.

3) All chemicals, apparatus, and trash must be removed from your desk before leaving the laboratory for the day. Any solid chemicals or solutions that are saved must be in containers fully labeled with your name, the chemicals' identity, and the date.

4) Your desk must be closed and locked before leaving for the day.

5) Use the laboratory's balances only in the correct manner (as described later in this manual).

6) Don't leave solid materials of any kind in the sink.
7) Don't attempt any experiments other than those described in this manual.

8) Don't return unused reagents to the stock bottle; we'll tell you how to dispose of them properly.

9) Obtain liquids in a beaker or bottle—don't insert your pipet or eyedropper into a stock bottle. Obtain solids in a beaker or on a watch glass. Replace the cap or stopper on the stock bottle *immediately*.

10) Don't heat bottles or volumetric glassware. The bottles are likely to break, and the volumetric flasks or glassware may break and will certainly be permanently distorted from their calibrated values.

11) Don't dry glassware with compressed air. It's dirty.

12) Record data promptly, but only in the lab notebook. Notes or data on loose paper will be confiscated and destroyed—obviously this carries its own punishment.

SAFETY PRECAUTIONS:

1) "Think bad in the lab!" Before you set out to do anything, think about ways it could go wrong and accidents that could happen, even unlikely ones—and then think about the precautions that should be taken to prevent trouble. Remember, Murphy's Law is waiting to nail you.

2) Notify the lab instructor *promptly* if any accident occurs in the laboratory, no matter how minor the accident seems to be. You may need more help than you think (and it *won't* hurt your grade). If you need to go to the Health Center, someone *must* accompany you!

3) Observe the location of the nearest fire extinguisher, the safety shower, the eyewash fountain, the first-aid instruction poster, and the lab exits.

4) Don't point the mouth of a test tube or flask at yourself or at a neighbor when heating substances. A suddenly formed bubble of vapor may eject the contents of the vessel violently and dangerously.

5) Don't add water to acid. Slowly pour the acid into water if the acid needs to be diluted.

6) Never taste a chemical. If it were food, you'd know it already. If you do get a chemical in your mouth, **don't swallow** — rinse your mouth out with lots of water without swallowing, and see the instructor immediately.

7) To observe the odor of any chemical, fan a little of the vapor toward you by sweeping your hand over the top of the container — **don't sniff it.**

8) Never use your mouth for suction to pipet liquids—use the pipet bulb in your lab desk.

9) Use a fume hood for all experiments involving poisonous or objectionable vapors or gases.

10) Wash your hands thoroughly with soap and water after leaving lab for the day and before eating.

ACCIDENT PROCEDURES:

CALL THE INSTRUCTOR IMMEDIATELY

Cuts: Allow small cuts to bleed freely and wash with cold water. The instructor can apply a Band-Aid. Let the instructor get you to medical care immediately for a major cut or severe bleeding.
Burns: Immediately place burned area under cold running water, then call the instructor to you.
Acid or base burns: Wash with large volumes of water and report to instructor.
Acid on clothing: Douse with dilute aqueous ammonia.
Acid on desk or floor: Add solid sodium carbonate or bicarbonate; when fizzing has subsided, flush with water.
Base on desk or floor: Add acetic acid, then flush with water.

NOTE THE FIRST AID INSTRUCTION POSTERS IN EACH LABORATORY!

If the accident results in injury to you or to anyone else, be sure that the instructor fills out or helps you fill out an accident report for the College's records.

WEIGHT AND BALANCE INSTRUCTIONS

MASS AND WEIGHT:

One of the most fundamental measurements of an object's physical properties is the measurement of its mass. The balance is a device for measuring the attraction of this mass for the mass of the earth. This attraction is gravitational force which we conveniently call *weight.* Since weight is proportional to the mass of the object being studied, and since it is so conveniently measured by means of a balance, we routinely make quantitative measurements on chemical systems by measuring weight, although mass is the fundamental property in which we are interested.

Traditional balances have done just what the name suggests: they have determined mass by comparing an unknown mass with another known mass, balancing the two against each other. Modern digital balances use an electric current to produce a force that opposes the weight of the unknown object, and read out the size of the current that is required to get a perfect balance of the two forces.

This laboratory has two kinds of balances, top-loading balances and analytical balances. The ***top-loading balance*** is convenient to use for weighings which need not be more accurate than about ±0.01 gram, and is quite sturdy. The ***analytical balance*** is accurate to about ±0.0001 gram, but the increase in sensitivity brings with it increased fragility. These analytical balances must be used with care! Since there are about 28 grams in an ounce, the intrinsic error in an analytical balance amounts to no more than four millionths of an ounce. Weight is one of the properties which can be measured most accurately, and you will want to take advantage of this fact in designing your experiments. Be sure in all cases, however, that you are using the appropriate balance for the task.

WHICH BALANCE TO USE?

The two general chemistry laboratories each have two Sartorius top-loading balances. This balance has a capacity of 200 grams and an accuracy of ±0.02 g. In general, it should be used for weighing chemicals, particularly for a chemical synthesis. As the instructions below will indicate, it is particularly easy to use — but it is less accurate than the analytical balances (Sartorius BA110), which each lab also has two of. Because the analytical balances are subject to corrosion and are fragile, **they should not be used for weighing chemicals for a synthesis** (in synthesis, the extra accuracy simply doesn't matter that much). A good rule to follow about balance use is:

It's OK to weigh chemicals ***onto*** a top-loader, but

Only weigh chemicals ***from*** an analytical balance!

This means that in chemical analysis (as opposed to synthesis), where great accuracy is absolutely necessary in weighing chemicals, the way to do it is to weigh by difference—fill a weighing bottle away from the balance, weigh it on the balance, then deliver a small sample from the weighing bottle to another container such as a beaker or flask and weigh the weighing bottle again; the weight difference between the first and second weighings is the weight of the sample you measured out. This is important! So, again: use a top-loader for weighing chemicals for a synthesis or for making up a solution that only has to be roughly correct, within a per cent or so. Use an analytical balance for very accurate weighings of samples for analysis, but don't ever pour the chemical sample into a container while it's on the balance pan.

USE OF THE TOP-LOADING BALANCE:

1) Make sure the balance is clean; if not, brush off the dry material, or use a Kimwipe to gently wipe off any wet residue. The pan *must* be clean and dry before beginning a weighing.

2) The display should be on and reading "**0.00**". If it's on but reads other than zero, press the ZERO button to get a zero reading. If it's blank, press the **ON** button; initially all the display segments will show: **8.8.8.8**, but in a couple of seconds it should read **"0.00 g"**.

[All the balances should have been calibrated by the assistants at the beginning of lab, but if there's been a power failure you'll need to recalibrate it. To calibrate, press the **CAL** button and hold it in until the display reads **"200.00"**. Place two 100-gram weights from the back of the balance in the center of the pan. Press the ENTER button; in a few seconds the display will read **"200.00"**. Remove the weights, allow the display to stabilize, and press the ZERO button to get **"0.00"**.]

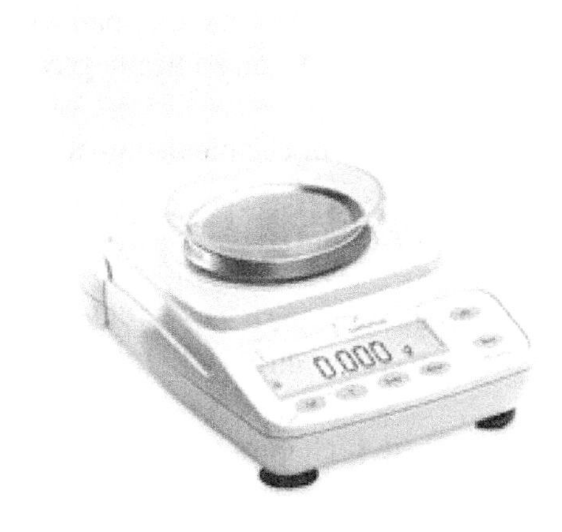

3) Place the appropriate container on the center of the pan: a beaker, a flask, or a piece of glazed weighing paper.

4) Press the **TARE** button. The weight of the container, previously visible, will disappear and **"0.00"** will appear again.

5) Add the desired chemical to the container until the appropriate weight is reached. Allow the final weight to stabilize for a few seconds. ***Record the weight in your notebook immediately.*** Remove the container and sample.

6) Wipe off any spills immediately.

USE OF THE ANALYTICAL BALANCE:

General Rules:

1) Don't weigh chemicals for syntheses on the analytical balance; that's not what it's for and you're wasting your own time!

2) The balance must be kept spotlessly clean!

3) The sliding doors must be kept closed when the balance is not being used, to prevent dust collection and corrosion. They must also be closed while weighing, because air currents will cause error in the indicated weight.

4) Don't ever weigh substances that produce corrosive fumes, such as acids.

5) Never weigh hot or warm objects. The upward current of warm air will make the object seem lighter than it really is.

6) Don't touch the balance pan with your fingers. Use finger cots or tongs to handle objects placed on the balance pan, so heavy greasy fingerprints don't cause a weighing error.

7) Never pour chemicals into a container sitting on the balance pan.

Weighing Procedure:

1) Make sure the balance is clean; if not, brush it off very gently with the brush attached to the balance. If it's very dirty, call the instructor.

2) **Close the doors after the balance is clean.**

3) Press the **T** (TARE) button. Wait for "0.0000" (five zeros) to appear.

4) Open the sliding door. Place the object to be weighed on the center of the pan. It should not hang over the sides—if it does, it's probably too heavy for the capacity of the balance!

5) **Close the door**. Wait for the weight to be displayed to four decimal places (only two dashes will show until the weight has stabilized). Don't open the door yet!

6) ***Record the weight indicated in your notebook***. Compare the notebook reading with the digital readout to be sure you haven't miscopied it.

7) Open the door and remove the object.

8) **Close the door!**

Weighing analytical samples by difference (see previous section on "Which balance to use?"). Do this whenever you have several samples to weigh out from a single container.

1) Away from the balance, fill your weighing bottle with an amount of the substance to be weighed that is *greater* than the total weight of the samples you expect to weigh. For three samples, each about 0.2000 grams, put in about 0.75 grams. If you're weighing out analytical samples of your synthesized metal complex, you can use the storage bottle directly instead of your weighing bottle.

2) Place the filled weighing bottle on the center of the pan and close the sliding door.

3) Wait for the weight of the weighing bottle to appear (indicating a stable weight) and press the **T** button. The digital readout should indicate **"0.0000"**.

4) Open the door, remove the weighing bottle, and pour a sample of the approximate desired size into another container on the balance table but away from the balance (for example, into a flask in which a titration is to be performed). Do *not* use a spatula or other tool to remove the substance from the weighing bottle; pour it out directly so that if the substance leaves the weighing bottle it has to be in the other container and can't have stuck to anything else. Replace the weighing bottle on the center of the balance pan. Close the door.

5) Wait for a stable reading. It will have a negative sign, indicating the amount you have *removed* from the weighing bottle. ***Copy the weight into your notebook*** (without the minus sign) as your sample weight. Compare the recorded weight to the digital readout to avoid error.

6) If the sample is much larger than you wanted, throw it out and start over. If it's smaller than you wanted, don't copy the weight down, just take the weighing bottle out and pour a little more into the other container, then check the weight again. Pretty soon you'll be in the right neighborhood.

7) When a sample of the desired weight has been achieved, record its weight and press **T**. The digital readout will read **"0.0000"** again.

8) Repeat steps (4) through (6) to get the weight of a second sample in a second container. (Remember to number the containers!) If you want to weigh out a third sample, press **T** again and repeat.

9) Open the door, remove the weighing bottle, and **close the door**. Make sure the balance is clean.

VOLUME AND GLASSWARE INSTRUCTIONS

Volume can't be measured with the same accuracy as weight. Consequently, careful volume measurements usually turn out to be measurements of the weight of a substance for which the weight of a given volume (the ***density***) is well known. This sort of procedure can be used to improve the accuracy of volumetric glassware. In our laboratory work we will be primarily interested in the volume of liquids. In measuring the volume of liquids, four devices are commonly used: the ***buret***, the ***pipet***, the ***volumetric flask,*** and the ***graduated cylinder***. Only the first three of these are sufficiently reproducible to make a weight calibration worthwhile.

The ***accuracy*** of these four devices is approximately as follows:

50 mL buret	±0.05 mL
25 mL analytical pipet	±0.03 mL
5 mL analytical pipet	±0.01 mL
500 mL volumetric flask	±0.10 mL
50 mL graduated cylinder	±0.25 mL
10 mL graduated cylinder	±0.10 mL

The ***precision*** (reproducibility) is somewhat better. It should be noted here that this degree of reproducibility can only be achieved if the various items of glassware are scrupulously clean. Water will uniformly wet clean glassware; if droplets remain on the walls of the glassware after draining, the glass is not clean and the volume will be quite inaccurate if the item of glassware is one which delivers a specified volume, as the pipet and buret do.

Dirty glassware can in many cases be cleaned with a brush and soapy water. If this doesn't work, or if a brush isn't appropriate (e.g., when cleaning a pipet), use the chromic acid cleaning solution. Place a few mL of this solution in the vessel, allow it to wet all the walls, pour the solution out, rinse with tap water and then with distilled water. Do *not* discard the cleaning solution unless it has turned green; if it is still orange, return it to the stock bottle. Chromic acid cleaning solution is prepared by dissolving sodium chromate in concentrated sulfuric acid. This solution vigorously attacks dirt and grease. It also vigorously attacks clothing and skin, if allowed to come in contact with them, so protect your hands with rubber gloves.

Deionized water is more expensive than you think. Three rinses of a few mL each are adequate for any item you will be rinsing.

Never heat volumetric glassware. Not even in the drying oven. It expands (changes volume) and doesn't contract back to the same size when it cools.

It is rarely necessary to dry volumetric glassware. It's usually better *not* to dry it, because drying almost always gets it dirty. If for some reason you must dry an item, rinse it with a few mL of ethanol (not acetone) and use an aspirator to pull air through it. Do NOT use compressed air; it's noisy and dirty.

Use of the Graduated Cylinder

Pour the liquid into the cylinder and allow about ten seconds for drainage. Read the volume from the scale on the side of the cylinder by sighting with your eye on the same level as the liquid surface. Almost all liquids wet clean glass surfaces; this means that the liquid surface in the cylinder will be, in general, concave (curving up at the rim). This curved liquid surface is called the ***meniscus***; always read the volume of the liquid as being that of the bottom of the meniscus, not that indicated by the higher edges. If a liquid is so strongly colored as to be opaque, or if it does not wet the glass (convex instead of concave meniscus), use the top of the meniscus, but not otherwise.

Use of the Pipet, Buret, and Volumetric Flask

After you have cleaned your volumetric flasks, they will be wet with distilled water inside. Leave them that way. You'll want to dry the outside, so you can be sure there aren't any water spots outside that you might later think were dirt on the inside—but store the volumetric flasks WET.

Store your burets WET. Once they are clean, fill them with distilled water and cork them. When you wish to use a buret, pour out the water, rinse with two or three 5-mL portions of your titrant solution, and then fill with the titrant. When you have finished with a buret for the day, drain it, rinse twice with 5-mL portions of distilled water, refill with distilled water, and cork.

Also leave the inside of your pipets wet. When you are ready to pipet a solution, rinse the pipet with a few portions (2-3 mL each) of the solution you will pipet. Your pipet will then be contaminated, but with the very solution that you wish to pipet. When finished, *rinse the pipet with distilled water* and leave the inside wet.

The volumetric flask is used in the following manner: Dissolve the solid sample in about half the volume of distilled water that will fill the volumetric flask, *using a clean beaker* (not the flask) and stirring or heating as needed. Transfer the solution quantitatively to the flask, rinsing the beaker at least three times (with distilled water) into the flask. Fill the flask to the bottom of the neck with distilled water and insert the stopper. Grasp the flask by the neck *with your thumb on the stopper* and invert it 4 or 5 times to mix the solution. Then remove the stopper and carefully add water (a wash bottle is convenient here) until the meniscus coincides with the mark on the neck of the flask. Reinsert the stopper and invert the flask at least 10 times to thoroughly mix the solution. ***NOTE:*** If the solution becomes either warm or cool when the sample material dissolves, wait for the solution to reach room temperature before diluting to the mark.

The correct method of using a pipet follows: With the pipet filled to about an inch above the mark and liquid held in place by the bulb suction, use a Kimwipe or clean paper towel to wipe off the wet area near the tip. Holding the pipet in a *vertical* position, allow the solution to drain out until the bottom of the meniscus is at the mark. Touch off the drop from the tip, and bring the pipet over the flask into which the delivery is to be made. With the pipet vertical, with the flask tilted about 30 degrees from the vertical, and with the tip of the pipet touching the wall of the flask, let the solution run into the flask. When the solution stops running out of the pipet, wait about 1 second with the pipet tip still touching the flask wall, and then remove the pipet. Do NOT blow out the drop of solution remaining in the pipet.

With a little practice, you should be able to operate the buret with your left hand while swirling the titration flask with your right. Working in this fashion, you can titrate much more rapidly than you can if you use just one hand for both stopcock and flask.

With Teflon stopcocks, as in the case of your burets, the easiest way to deliver a small volume of liquid (e.g., near the end point of a titration) is to turn the stopcock rapidly past the open position. Be careful, however; overly vigorous stopcock twisting can break the buret.

An ***erlenmeyer flask*** is the vessel of choice for titrations. Limit the quantity of solutions so that the flask is not filled more than half way. After each addition of titrant to the flask, swirl it to mix in the titrant. When you are quite close to the end point, rinse down the walls of the flask with a few mL of distilled water from your squirt bottle. Then continue with the titration, one drop at a time.

Adjust your sample size for any titration so that about 35-40 mL of titrant will be required.

You should note that the volumetric glassware has a temperature, almost always 20°C, indicated on it. This is the temperature at which the manufacturer intends his calibration to apply. If the liquid to be measured out is at some other temperature, the calibration will be different because glass and water don't expand at the same rate as temperature increases. Corrections can be made for this effect, but it's usually not necessary if the liquid's temperature is within about five degrees of the temperature at which you or the manufacturer calibrated the glassware.

SOLUTION CONCENTRATIONS AND PREPARATION

The two physical properties we've just described, weight and volume, are the most basic physical measurements a chemist usually makes. In a general way, it's obvious that these are fundamental ways to measure the quantity of matter that we're working with. However, it frequently turns out that we need the *ratio* of these two quantities, in that we need to know the weight of one substance (the ***solute***) dissolved in a given volume of a liquid (the ***solvent***), forming a ***solution***.

Solution concentrations come in a variety of definitions, not all of which involve both weight and volume. For example, in biological applications, solutions are frequently made up on a percent composition basis. This can involve a weight percentage, in which the percentage refers to the weight of solute dissolved in each 100 grams of total solution weight; thus a solution of 3 grams of NaCl in 97 grams of water (enough to make the whole solution weigh 100 grams) would be labeled "**3% w/w**", meaning 3% weight-for-weight. When the solute is also a liquid, as for example ethanol, it's fairly common to express the percentage on a volume-for-volume basis. A solution of 30 mL ethanol in 70 mL water would be labeled "**30% v/v**", meaning that the solute was made up to have the solute ethanol occupy 30% of the total volume of the solution, if the volumes really do add linearly. In this case the volume of the solution would be a bit smaller than 100 mL, maybe 97 mL, but this sort of concentration measurement is not normally used when the concentration must be known very exactly. In the applications for which we will be using solutions, all or nearly all of the solutions will be made up on a ***molar*** basis. The definition of molar concentration is

$$\frac{x \text{ moles solute}}{1.000 \text{ L solution}} \equiv x \text{ M} \quad (\textit{molar concentration})$$

This definition should make it clear that in order to make up a solution of known molar concentration (or ***molarity***) we'll need to measure both mass (in order to calculate moles of solute) and volume (in order to get liters of solution). So we'll expect in general to weigh out a given amount of solute on a balance, so that (with its chemical formula and molecular weight) we can calculate how many moles of solute are present. Then we'll dissolve it in a volume of solute somewhat smaller than the target volume in liters, and dilute it up to the desired volume in the appropriate piece of glassware. Of course, there's no law saying that you have to make up a volume of 1.000 L just because the molarity definition is quoted as "per liter". A 0.6 M solution *can* be made up by dissolving 0.6 moles of solute in a volume of 1 liter, but it can just as well be made up by dissolving 0.3 moles in 0.5 L, or by dissolving 0.06 moles in 100 mL—or any other pair of numbers that make the ratio come out right.

Quick and Dirty Solutions: There are two different levels of care that are needed in preparing solutions at a given molar concentration, depending on what you expect to do with the solution when you get it. Some solutions only need to be made up on a rather loose approximate basis, either because we're always going to use an excess of them anyway or because the purity of the solute can't be guaranteed (so that in effect we don't *really* know the molecular weight of the solute). For this kind of solution, it's adequate to use the top-loading balance that only weighs to 0.01 g, and it's satisfactory to measure volume equally casually, either by using the marks on the side of a beaker or by simply filling a screw-cap bottle of the right standard size to the top. Of course, when you've done this you don't know the molarity to any very good accuracy, and it should only be quoted to one or two significant figures: 0.012 M, or 0.5 M.

Careful Solutions: On the other hand, some molar solutions are going to be used as fundamental molar reference amounts (called ***primary standards***). For such a solution, the solute and solvent will have to be very pure, and the amounts will have to be measured with the greatest possible accuracy. To get the weight, we would expect to use the analytical balance, weighing to an accuracy of ±0.0002 gram. Then to measure the volume of the solution we would use a volumetric flask, measuring volume to (in general) about ±0.1 mL. (Very small volumetrics are more accurate than this, and large ones are less accurate—but each will be accurate to about ±0.5% of the total volume it contains.) Such a solution should have its concentration quoted to four significant figures: 1.036 M, 0.1193 M, or 0.02269 M. (Notice in that last concentration the only figures that count are the ones after you get through using zeros to prop up the decimal—another way to write it would be 2.269×10^{-2} M, which pretty clearly has exactly four significant figures.)

We ought to note that even for a primary standard solution, it's possible to use approximate weights if you use them in the right way. "The right way" means that although we need to know four significant figures for the concentration, it doesn't need to be a precise *round* number. Later in the semester, you'll need to make up a standard zinc solution that's near 0.02 M; notice that that could mean 0.02000 M, but it could just as well mean 0.01838 M, or 0.02243 M, or anywhere else within about 10% of the target 0.02. It's just that *whatever it is*, we need to know four significant figures. In weighing the zinc metal that will be used, you'll find that the target weight is 0.6538 g. Of course you'll need the analytical balance, but you *don't* need to try to shave teeny pieces of zinc off the metal to get exactly 0.6538 grams; instead, get within about 10% of that—somewhere between 0.55 and 0.70 g, but weighed to four decimal places. 0.5986 g is fine, and so is 0.6989 g or 0.5351 g.

How to do the Necessary Calculations: Let's look at an example. Suppose we want to make up a standard solution of Co^{2+} [cobalt(II) ion] at about 0.05 M, and we're going to do titrations with the solution; each titration will use up about 50 mL of the solution, and we'll need to do somewhere between three and six titrations. How much solution should we make up, and how should we make it up? For a standard solution, the volume will have to be measured in a volumetric flask, and those come in round-number sizes: 10 mL, 25 mL, 50 mL, 100 mL, 250 mL, 500 mL, 1 L, 2 L. Three titrations would use up about 150 mL, but six would use up 300 mL. It's better to be safe by making up a bit more solution than you think you're going to need, so as not to have to make it up twice. So we should choose a 500 mL volumetric flask, which will give us plenty. Now what cobalt chemical are we going to weigh out, and how much should we shoot for? It would be necessary to go to a reference book for analytical chemistry at this point, in order to find out what the primary standard cobalt substance is. Here we'll tell you that the material to use is cobalt metal—the element. It will need to be dissolved in nitric acid (HNO_3). Cobalt metal is available in very small granules, so we can presumably get pretty close to a target weight without trouble; how much do we need?

We can get at the weight of cobalt metal by doing a little dimensional analysis calculation. What we want to know is how many grams of cobalt metal are required; our target is the 0.05 M solution that we have to wind up with. Then we can convert moles of cobalt ion per liter to moles of cobalt ion to moles of cobalt metal to grams of cobalt metal:

$$? \ g\ Co \equiv \frac{0.05\, moles\, Co^{2+}}{1L\ soln} \times 0.5L\ soln \times \frac{1\, mole\, Co}{1\, mole\, Co^{2+}} \times \frac{58.93\, g\, Co}{1\, mole\, Co} = 1.4733\ g\ Co$$

How to Make Up the Solution: So our target weight of cobalt metal is 1.4733 grams. To get the cobalt metal, you would normally take a snap-cap vial to the stockroom, labeled **"Co metal, 1.5 g"**. You'd get a little extra, and the way to get an accurate weight would be to take a 600-mL beaker and the vial of Co to an analytical balance, along with your notebook to write down the weight. Tare the balance at 0.0000, place the vial on it, and let it stabilize at the total weight of the vial plus the Co metal, say 14.9893 g. Then press

T again to get 0.0000 as the *relative* weight of the filled vial. Remove the vial from the balance, pour what looks like about 1.47 g of the cobalt metal into the beaker, and reweigh the vial. Now the balance will read a negative number, say -1.3956 g, which would mean that you had transferred 1.3956 g of Co metal into the beaker. That's close enough to 1.4733 g—all you need is to get within about 10% of 1.47, say between 1.30 and 1.60 g.

Now you would need to dissolve the cobalt metal in nitric acid, which should be done in the fume hood. It would only take perhaps 10 or 15 mL of nitric acid, with perhaps a little distilled water to moderate the reaction. When the metal has dissolved, use distilled water (the pure solvent principle) to dilute the solution up to about 400 mL—remember that we're going to use a 500 mL volumetric flask, so stay well inside that volume. Pour the cobalt solution into the volumetric, and rinse the beaker three times into the volumetric with about 10 mL distilled water, so as to be sure all the cobalt solution has been transferred. Now use distilled water from a small beaker to fill the volumetric flask up to about half an inch below the fill line on the neck of the flask, then an eyedropper to add the last little bit, until the *bottom* of the liquid surface (the ***meniscus***) in the neck of the flask is just even with the line. Stopper the volumetric flask, hold the stopper in place, and invert the flask 20 times to be sure that you've thoroughly mixed the solution in the flask. Check the meniscus level again to be sure mixing hasn't changed the volume of the solution.

What Did You Really Get? At this point you've prepared a solution of Co^{2+} whose molar concentration is known to four significant figures, because you know the weight of cobalt metal to five significant figures (1.3956 g) and the volume of the solution to four significant figures (500.0 ± 0.1 mL). Exactly what is the molar concentration? Again, use dimensional analysis to have the units do the calculation for you:

$$? \frac{mol\ Co^{2+}}{L\,soln} = \frac{1.3956\,g\,Co}{0.5000\,L\,soln} \times \frac{1\,mol\,Co}{58.93\,g\,Co} \times \frac{1\,mol\,Co^{2+}}{1\,mol\,Co} = 0.04736\ M$$

Now it's true that we wanted to make up a solution that was 0.05 M, but this is a perfectly satisfactory number; 0.04736 is certainly within 10% of 0.05, which is a reasonable rule of thumb. Notice that the important point is *not* to have the concentration be exactly 0.05000 M, but to have it near there and to *know* what it is to four significant figures!

EXPERIMENTAL ERRORS AND PRECISION

Accuracy and Precision

Before making any measurements one should be able to distinguish carefully between the two terms accuracy and precision. ***Accuracy*** is a measure of how closely the measured value agrees with the true or accepted value. ***Precision*** is a measure of how closely a series of measurements of the same quantity agree with each other. Note that an experiment may be precise but not accurate, just as three bullets fired at a target may strike the target very close to each other but still far away from the bullseye. Good precision in an experimental measurement is obtained by repeating the experiment over and over again until consistent results are obtained. In actual research, sometimes hundreds or even thousands of trials are run in order to obtain the desired precision. In this laboratory work in general chemistry you will be expected to repeat all analytical determinations until you obtain a minimum of three trials that agree with one another within the expected precision of the overall measurement. You can use statistics to tell you when you're there; we'll get to that.

In planning experiments, use instruments and techniques that have approximately the same precision. Using a more precise measurement for only one variable will not improve the overall precision of the experiment, which is determined by the weakest link in the chain of measurements. The overall precision for most of the apparatus you will use is shown below. (Note: The third column shows how you should record typical measurements in your notebook.

Instrument	Precision	Typical Notebook Record
Top-loading balance	0.01 g	11.24 ± 0.01 g
Analytical balance	0.0002 g	1.0200 ± 0.0002 g
50 mL buret	0.05 mL	25.42 ± 0.05 mL
25 mL analytical pipet	0.03 mL	25.00 ± 0.03 mL
5 mL analytical pipet	0.01 mL	10.00 ± 0.01 mL
50 mL graduated cylinder	0.2 mL	10.0 ± 0.2 mL
10 mL graduated cylinder	0.1 mL	6.3 ± 0.1 mL
-10/+150°C thermometer	0.3°C	50.3 ± 0.3°C

Whenever you have a measurement to make as part of an analytical project—such as the titrations in your analysis, for instance—try to set up the measured quantity in such a way that the intrinsic measurement error in the equipment (as indicated in the table above) is not more than about 1/1000 of the measured amount. Most analytical methods can be carried out to this accuracy. For a weighing on the analytical balance, this means that since the intrinsic error of the balance is 0.0002 g, the sample being weighed should weigh at least 0.2 g (1000 × error). For a volume measurement with a buret, since the error is 0.05 ml the volume to be used should be close to 50.00 mL (1000 × error); however, to allow a little safety margin in this case since 50.00 mL is the limit of measurement with the buret it's better to calculate quantities for about 40 mL delivered by the buret. This sort of consideration can be used to decide how big samples need to be for analysis, and we expect you to do this in planning any analysis.

Determination of Overall Accuracy—Percent Error

When the actual or expected value is known, the overall accuracy can be expressed as ***% error***. Percent error is calculated as follows:

$$\%\ error = \frac{Experimental\ value - Accepted\ value}{Accepted\ value} \times 100$$

For example:

Accepted percent zinc in a coordination compound is 24.49%. Measured or experimental value is 24.77%.

$$\%\ error = \frac{24.77 - 24.49}{24.49} \times 100 = \frac{0.28}{24.49} \times 100 = +\ 1.14\%$$

Here the measured value is larger than the accepted value; this is indicated by the "+" sign on the % error. If the measured value had been 1% low, it would have been expressed as -1.0%.

Significant Figures

In recording experimental data it is customary to indicate the accuracy of the measurement by using a certain number of figures or digits, called significant figures. The number of significant figures (or in faculty shorthand ***sig figs***) in a quantity is the number of trustworthy figures—the last figure, however, being somewhat doubtful within the precision of the measurement. Suppose an object was weighed on a balance sensitive to 0.02 g and the weight was found to be 25.93 g. This result has four significant figures within an accuracy of 0.02 g. In other words, we know its weight lies between 25.91 and 25.95 g; to indicate this properly we would write 25.93 ± 0.02 g. Note that the *last significant figure is uncertain*. Examples of the number of significant figures in various numbers are seen in the table below.

Number	Significant Figures
0.49	2
6.93	3
20.5	3
0.000495	3 (*not* 6!)
694035.	6
20.0	3

Note that zero may or may not be a significant figure depending on its location. If a zero occurs between other digits, as in 694035, or it is the only number to the right of the decimal point, as in 20.0, it *is* a significant figure. If a zero appears before other digits, as in 0.000495, it merely serves to fix the decimal point and *is not* a significant figure. If a zero appears after other digits it may or may not be significant. Suppose an object is weighed on an analytical balance to the nearest 0.0002 g., and is found to weigh 2.3610 g. The zero here is significant since it is within the accuracy of the measuring instrument. However, if the weight from this same balance had been reported as 2.361000 g., the last two zeros are beyond the accuracy of the balance and would not be significant.

The advent of small electronic calculators has made laboratory calculations a lot more convenient, but they have some deficiencies as far as significant figures are concerned. There are two basic problems:

(1) If your calculation yields a long decimal number, the calculator is going to display its maximum number of digits—8, 10, 12—regardless of their significance .

(2) If your calculation comes out even by cancellation (for example, 3.3986 - 1.3986 = 2.0000), the calculator will truncate the trailing zeros (it'll read 2.) even though they're significant. In either case, you have to think about the true level of significance of the calculated result and either cut the number of digits down or add on zeros as needed.

Our principle is this—the calculator is quicker than you are, but you're smarter than it is. You're responsible for being smarter!

In addition, subtraction, multiplication, or division of numbers involving different numbers of significant figures, the final answer cannot have greater accuracy than the least accurate number used, as illustrated in the following examples.

Addition	Multiplication
2.0 32	23.15 x 1.623 = 37.55622
21.2	= 37.56
15.6 3	
38.8 62 = 38.9	

When numbers are to be rounded off, if the number to be dropped is greater than 5, the next higher number is used for the last significant figure. If the number to be dropped is less than 5, the last retained number is left unchanged. When the number to be dropped is 5, round off the last retained digit to an even number (i.e., if the figure before the 5 is odd increase the figure by one, but if it is even leave the figure unchanged).

Recording Errors and Significant Figures in Your Notebook

In recording all numerical data in your notebook be sure to ***(1) show the units involved*** and ***(2) use the appropriate number of significant figures***. Where applicable, calculate the average and the standard deviation for any overall series of measurements. ***NOTEBOOKS WILL BE GRADED OFF SEVERELY IF YOU FAIL TO FOLLOW THESE PRACTICES.***

The Assessment of Reproducibility (Precision)

If many measurements of a single quantity are made, even with good technique, they won't all yield the same number. Of course, the average is taken as the best value in this case. However, another quantity of interest to us is the amount of scatter in the numbers. 1.028 is much more reliable to us if it's the average of 1.027 and 1.029 than if it's the average of 0.578 and 1.478. One way to evaluate the scatter (or, thinking positively, the precision) is to get the average deviation:

$$\text{deviation of nth measurement} = d_n = |\langle X \rangle - X_n| ,$$

where $\langle X \rangle$ is the average measurement, X_n is the *n*th measurement, and $|\ |$ indicates the absolute value (plus sign).

$$\text{average deviation} = \langle d \rangle = \frac{d_1 + d_2 + \ldots + d_n}{n}$$

(if a total of *n* values were measured)

However, if many measurements are made and plotted as a bar graph or ***histogram*** for narrow ranges of values around the true value:

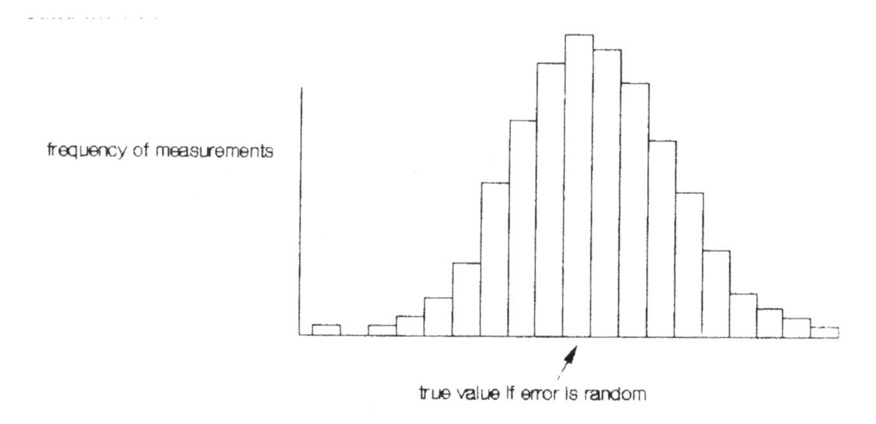

the function that results, called the ***error function***, is better described by the distance out to the steepest part of the curve. This quantity is known as the ***standard deviation***:

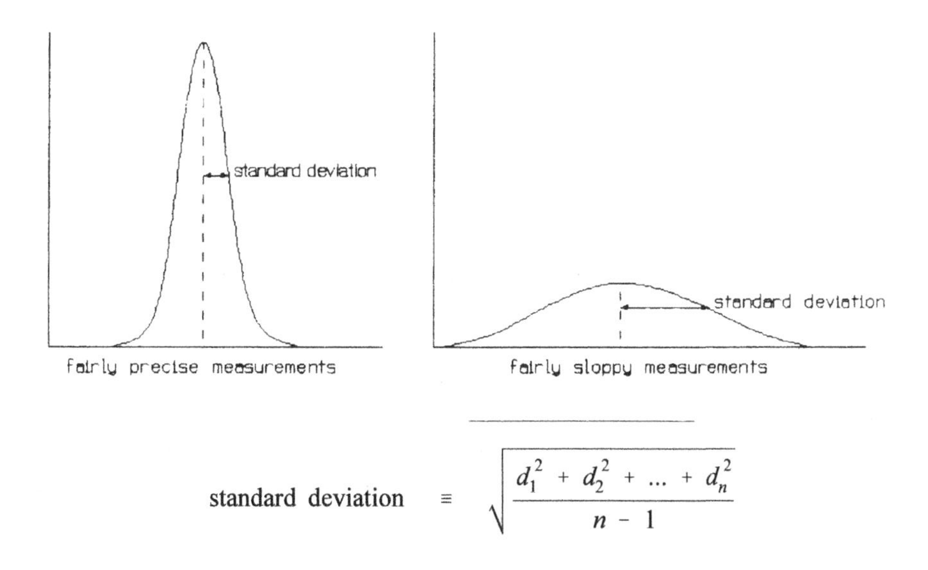

$$\textbf{standard deviation} \equiv \sqrt{\frac{d_1^2 + d_2^2 + \ldots + d_n^2}{n - 1}}$$

At several points in the laboratory work when you have made careful measurements an assessment of precision will be in order. Usually the best quantity to use will be this standard deviation.

Well, how good is good? If you have a standard deviation (call it σ, Greek *sigma*), does it correspond to the "fairly precise measurements" graph above or to the "fairly sloppy measurements" graph? The only way to tell is to compare σ to the mean value. That is, if the mean value is 0.1282 a standard deviation of 0.0626 is pretty awful — it's half the mean value — but if the mean value is 6318.2925 a standard deviation of 0.0626 is magnificent in the sense that it's tiny compared to the mean. You could quote σ as a percentage of the mean, but careful chemical measurements usually use ***parts per thousand*** instead, which is just like percentage (where you wind up multiplying by 100) except that you multiply by 1000 instead. A good set of chemical measurements will have a standard deviation that's only about *one part per thousand* of the mean (written as *1 ppt*). A set of measurements with σ at about 3 ppt is probably OK, but a set with σ at 8 or 9 ppt is pretty sloppy and needs to be improved. The standard deviation expressed as parts per thousand of the mean is called the ***relative standard deviation***, calculated as

$$\text{relative standard deviation} = \frac{\text{standard deviation s}}{\text{mean value } \langle X \rangle} \times 1000$$

Let's look at an example.

A solution is known to have a concentration somewhere around 0.02 moles per liter, or 0.02 M. Two students analyze the solution to get an accurate concentration. One student, Sydnor Tidy, does the analysis four times and gets the concentration values 0.01891 M, 0.01887 M, 0.01889 M, and 0.01892 M. When he uses a calculator to get the mean of these values, $\langle X \rangle = 0.0188975$ M (from the calculator readout) and $\sigma = 0.000022174$ M. Based on these values the relative standard deviation, calculated as above, is 1.2 ppt, and that's a good set of measurements. But those aren't the numbers he should write in his notebook as final data, because they both quote too many figures — not all the figures are significant figures. The standard way to report data is to use the mean to as many decimal places as it takes to get to a significant figure in the standard deviation. Here that means reporting the mean to five decimals as 0.01890 (rounding off) because it's the fifth decimal place in σ before you get to a nonzero number. Then you report σ to two sig figs (one decimal place past the end of the sig figs in the mean value), which here would mean $\sigma = 0.000022$. The compact way to write this in a notebook is: "$\langle \text{concentration} \rangle = 0.01890 \pm 0.000022$ M, 1.2 ppt".

The second student, Irv Gratch, gets 0.01886 M, 0.01861 M, 0.01890 M, and 0.01891 M from his four trials. He uses a calculator on these and gets $\langle X \rangle = 0.01882000$ and $\sigma = 0.000141657$. These two values yield a relative standard deviation of 7.5 ppt, which is fairly sloppy. He *could* run some more trials to try to improve his values, but let's look at *why* his first four trials are sloppy. His values are all around 0.01890 except for the second one, which is way down at 0.01861. That value is suspicious — maybe he spilled a bit of solution or something without realizing it. Now, the standard deviation is only intended to describe the spread of random error, not to cover a blunder like spilling solution. If Irv could be sure that second value was a blunder, he ought to throw it out and do the statistics only on the three good values.

If Irv actually *saw* the solution spill, he should quit right there and note in his notebook that the trial was discarded because of a spill. But if he didn't see anything amiss, it's still possible to use statistical tests to throw out irregular values. A convenient statistical measure for discarding a suspicious value is the ***Q test***. The Q test is based on the idea that the farther a measurement is from the mean, the more likely it is that the measurement is a fluke. What you do is to set up a ratio or quotient (hence the *Q*) of (1) the distance between the funny result and its nearest neighbor to (2) the total range from top to bottom of all the measured results:

$$Q = \frac{X_{suspicious} - X_{nearest\,neighbor}}{X_{suspicious} - X_{farthest\,away}}$$

This assumes, of course, that the suspicious result is way out on one end of your measurements; Q, if you think about it, can't get any bigger than 1.000. Now, to be 90% sure that the suspicious result ought to be rejected, discard it if its Q value is greater than the rejection quotient $Q_{0.90}$ (for the number of trials you've made) in the table on the next page.

$Q_{0.90}$ Values for Various Numbers of Trials

(Rejection Quotients at 90% Confidence)

Number of Trials	$Q_{0.90}$
3	0.941
4	0.765
5	0.642
6	0.560
7	0.507
8	0.479
9	0.441
10	0.409

In Irv's case, he sets up the Q fraction this way:

$$Q = \frac{0.01861 - 0.01886}{0.01861 - 0.01891} = 0.8333$$

He goes to the table above and finds that for four trials the $Q_{0.90}$ value is 0.765. Since his data yield a Q value of 0.833, which is *bigger* than the 0.90 or 90% Q value in the table, he can be at least 90% sure that the suspicious value 0.01861 is not a legitimate experimental error but a blunder and ought to be thrown out. So he throws out that value, which leaves him with three legitimate values, and he goes back to the calculator and does the statistics over again on the three good values: $\langle X \rangle = 0.0188900$, $\sigma = 0.000026458$. These numbers yield a relative standard deviation of 1.4 ppt, which is now quite good. Using the standard notation for his values, he reports "$\langle$concentration$\rangle = 0.01889 \pm 0.000026$, 1.4 ppt".

Summarizing, our two students' results look like this:

	Tidy	*Gratch original*	*Gratch after rejection*
Trial 1	0.01891	0.01886	0.01886
Trial 2	0.01887	0.01861	0.01890
Trial 3	0.01889	0.01890	0.01891
Trial 4	0.01892	0.01891	
Mean	0.01889 750	0.01882 00	0.01889 00
Std dev	0.00002 217	0.00014 1657	0.00002 6458
Reported mean	0.01890	0.01882	0.01889
Reported std dev	0.000022	0.000142	0.000026
PPT	1.2	7.5	1.4

There's one more bit of statistics you'll need. The standard deviation σ that we've described essentially tells you the answer to the following question: "If I do one more determination, how far will that result probably be from the mean value?" A more important question is: "How close is the mean to the unknown true value?" The answer to that question is the ***standard deviation of the mean***, which is just the ordinary standard deviation divided by the square root of the number of trials:

$$\text{standard deviation of the mean} = \frac{\text{standard deviation } \sigma}{\sqrt{n}\ (\text{number of trials})}$$

Obviously, you can also get a ***relative standard deviation of the mean*** in *ppt*, which is just the standard deviation of the mean, divided by the mean value, times 1000:

$$\text{relative standard deviation of the mean} = \frac{\text{standard deviation } \sigma}{\sqrt{n} \times \text{mean } \langle X \rangle} \times 1000$$

How to use EXCEL spreadsheet to do statistics

1. Click on the EXCEL icon to open the program: you'll see an array of little boxes – "cells" – the cells are identified by the column they're in and then the row they're in: e.g., A1, A2, A3…

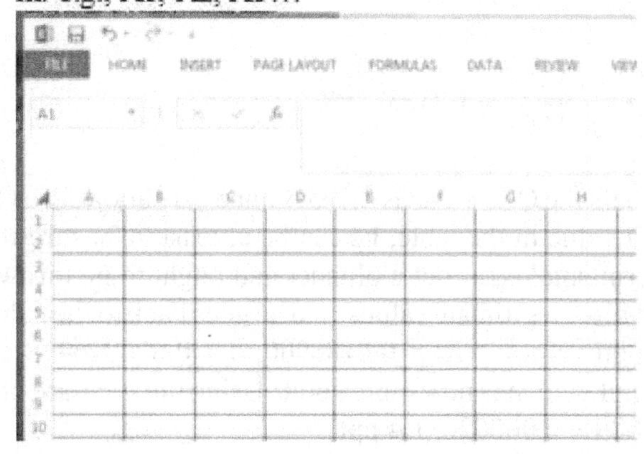

2. Enter your data in the cells of the first column: A1, A2, A3, and A4 (as on page 28):

	A	B
1	1.2207	
2	1.2213	
3	1.2199	
4	1.2240	
5		
6		
7		
8		
9		
10		
11		

If you have three data points, use the first three cells in the column. The extra cells may be used if you have more than four numbers to be treated (up to nine points).

3. Because you will sometimes have more than 3 or 4 numbers, skip down to A10 and write:
=AVERAGE(A1..A9).

	A	B
1	1.2207	
2	1.2213	
3	1.2199	
4	1.2240	
5		
6		
7		
8		
9		
10	=AVERAGE(A1..A9)	
11		

4. When you press the ENTER key on your keyboard, the result is 1.2215, which is the average of your three numbers.

	A	B
1	1.2207	
2	1.2213	
3	1.2199	
4	1.2240	
5		
6		
7		
8		
9		
10	1.2215	
11		

5. Now go to cell A11 and write: =STDEV(A1..A9).

	A	B
1	1.2207	
2	1.2213	
3	1.2199	
4	1.2240	
5		
6		
7		
8		
9		
10	1.2215	
11	=STDEV(A1:A9)	
12		

6. When you press ENTER, the results is the standard deviation of your three numbers, 0.0018.

	A	B
1	1.2207	
2	1.2213	
3	1.2199	
4	1.2240	
5		
6		
7		
8		
9		
10	1.2215	
11	0.0018	
12		

7. To calculate the standard deviation of the mean, you need to divide the standard deviation by the square root of the number of trials:

	A	B	C
1	1.2207		
2	1.2213		
3	1.2199		
4	1.2240		
5			
6			
7			
8			
9			
10	1.2215		
11	0.0018		
12	=A11/SQRT(COUNT(A1:A9))		

8. When you press ENTER, the standard deviation of the mean, 0.000889, appears. To express this as parts-per-thousand, multiply by 1000 (in your head) to get 0.889.

	A	B
1	1.2207	
2	1.2213	
3	1.2199	
4	1.2240	
5		
6		
7		
8		
9		
10	1.2215	
11	0.0018	
12	0.000889	

9. To calculate the **relative** standard deviation of the mean, you divide the standard deviation of the mean by the mean itself:

	A	B
1	1.2207	
2	1.2213	
3	1.2199	
4	1.2240	
5		
6		
7		
8		
9		
10	1.2215	
11	0.0018	
12	0.000889	
13	=A12/A10	

10. When you press Enter, the relative standard deviation of the mean, 0.000728, appears. To express this as parts-per-thousand, multiply by 1000 (in your head) to get 0.728.

	A	B	C	D
1	1.2207			
2	1.2213			
3	1.2199			
4	1.2240			
5				
6				
7				
8				
9				
10	1.2215		Mean	
11	0.0018		Std Dev	
12	0.000889		Std Dev Mean	
13	0.000728		Rel Std Dev Mean	

NOTEBOOK RECORD OF STATISTICAL CALCULATIONS

However you compute the statistics for your data - whether by pocket calculator, your own computer, or the chemistry department computers - you will want to record those calculations in your laboratory notebook. In doing that you'll need to be careful writing numbers to avoid transposing digits.

The chemistry department computers in Gilmer 215 and 216 are equipped with printers that can print your data and statistical results onto a label that can be pasted into your notebook. Next to each computer there is a laminated sheet of instructions that explains how to access EXCEL and how to print out your results. If you print labels paste them immediately into your notebook along with a descriptive title that identifies the results that have been calculated, e.g., "Stats for known Cobalt titrations," "Stats for $CoCl_2$-2PNO," etc., not just "Stats" or "My Stats."

Remember CARD - Computers are really dumb! If you put wrong data into your calculator or EXCEL, the results you get will also be wrong. So be sure to check the numbers that you enter for accuracy before you tell the computer to calculate. [Computer scientists dating back to the 1950s have used the acronym "**GIGO**" to describe that situation - **G**arbage **In** **G**arbage **Out**!]

Statistics and the TI-83 Keyboard

The TI-83 keyboard is shown below. The menu keys for doing statistics are the STAT PLOT key and the STAT key, Various submenus appear when you press these keys.

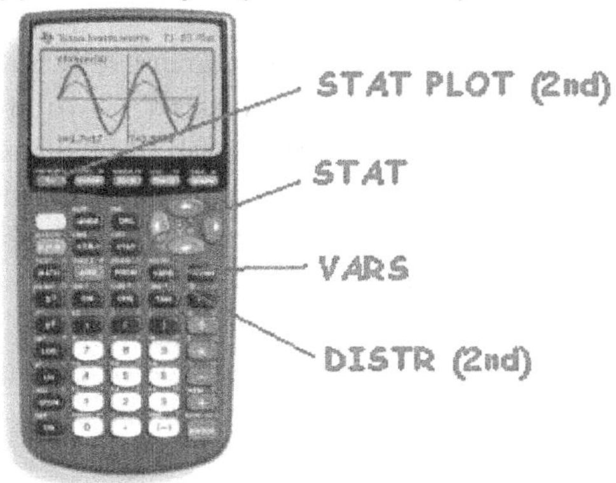

Entering Data:

- Statistical data are stored in the TI-83 as **lists**. Up to six lists can be stored.
- To enter data, press STAT and got to EDIT. *Don't press LIST.*
  - In the first empty column (usually column L1), enter the successive values for the variable whose statistics you want to find.

Finding the Descriptive Statistics:

Press STAT and activate CALC.

- press 1:1-Var Stats.
  - Select the variable for which you wish to do the computation. For example, to compute the descriptive statistics for the variable in L1, press "2nd 1" = L1 .
  - The screen should now read 1-Var Stats L1.
  - Press ENTER. Now the mean and standard deviation for your variable will appear on the screen. If you practice with the three values 1, 2, and 3, the screen displays:

1-Var Stats

$X = 2$ ← This is the mean of 1, 2, 3

$\sum X = 6$

$\sum X^2 = 14$

$S_X = 1$ ← This is the standard deviation of 1, 2, 3

$\sigma_X = 0.8164965809$

$n = 3$

LABORATORY NOTEBOOK INSTRUCTIONS

Introduction

The laboratory notebook is the fundamental working document of scientific discovery, and as such it is of extreme importance. The process of discovery is inductive—we are learning what the rules of the game are by watching the game—and this differs very sharply from a textbook and even from some laboratory manuals, which are written in such a way as to make us think deductively—we write the rules out ahead of time and make the game conform to them. For most of you, then, the laboratory notebook will be an entirely new type of construction in writing.

We have already emphasized, and properly so, the importance of preparing your laboratory work in advance. To complete the laboratory in the allotted time, you must know when you come into the laboratory what you plan to do. This means that you should have outlined in your notebook a stepwise plan of action, and should have allowed space for tables of data, etc. Unfortunately, a common error—and a grave one—is to presume that this constitutes the entire function of the notebook. It will very frequently happen that the stepwise plan of an action you had devised will not work. The solution will turn red when you expected a green solid to precipitate. What then?

This is where the notebook must be used properly, and *in a different way* from fill-in-the-blank lab manuals. It is true that the notebook should contain a plan for what you expect to happen in the lab. Where this differs from your expectation, the notebook should contain your interpretation of the differences—why they occurred, what they mean. The differences will lead you to modify your plan of attack, and the notebook should contain a summary of your revised approach. Repeating this process, and repeating the experiment, will eventually give you a result that can be interpreted within the scope of the most recent plan of action in the notebook. At that point the experiment has been completed, and it should be described in full, then interpreted.

What is important to realize is that the intermediate failures must *also* be recorded in full and interpreted. Only from these unpredictable developments can we learn how to refine and improve a technique, or how to derive information about a chemical system which we had previously not suspected. Your notebook is very basically improper if it is written up in advance and no further comment is added about what actually went on in the lab. Even on the occasional afternoons when everything goes as you had anticipated, you *must* comment as you go along on the progress of the experiment. We can only discover the rules of the game if we have a complete record of what happened as the game progressed.

Since your notebook is a complete record of all the things you did in the laboratory, it is the sole basis for judgment of your laboratory performance—that is, your grade is based totally on your notebook. Don't try to make yourself look smart by omitting the mistakes and errors of judgment—they will absolutely not count against you if you learned from them and didn't repeat them. What will count against you is an incomplete record. We can't give you credit for your effort if we don't know what you did, and the only way we have of knowing what you did is through reading your notebook. So plan it ahead, but keep it up to date as you go along. As we go along, we'll provide bulleted lists of headings you need in the notebook at that stage.

Structuring and Using Your Notebook

1. We'll give you a quadrille-ruled notebook with sewn-in pages; use only that kind. If you run out, get another one from the stockroom.

2. Before you use it in lab, start with the first *quadrille-ruled* page (not the flyleaf) and number the pages consecutively all the way through to the end. Put the numeral in the center of the top of each page. Left-hand pages should have even numbers, right-hand pages should have odd numbers. Make the numerals in ink, either blue or black.

3. Put your name inside the front cover, and label the opposite page (page 1) "TABLE OF CONTENTS" at the top.

4. Leave pages 1 through 4 for the table of contents. You almost certainly won't need this much room, but since you're not sure right now—and neither are we—it's best to allow plenty of room.

5. Start your writeup of actual lab procedures and results **on page 5**, a right-hand page. **In general, procedures should be described and data recorded on right-hand pages. Left-hand pages should be used for sketches, calculations, and scratch paper.** ALL notebook entries of any sort must be in blue or black ink, and you must *never* erase an entry or cover it with whiteout liquid! And ***NEVER*** remove a page from the notebook! A *complete* record is essential—that's why the pages are sewn in.

6. When you first use a right-hand page—then and only then—you should write your initials and last name in the upper right-hand corner, with the date of that page's work immediately under your name. This is an important validation procedure, and we'll grade your notebook down if it's not done.

7. Title page 5 "***SEMESTER PROJECT: SYNTHESIS AND ANALYSIS OF A COORDINATION COMPOUND***" Under that, put "**EXPERIMENT 1: Literature Search for Synthetic Procedure**". In the lab lecture, we'll give you some guidance as to what to do first.

8. You'll need some subheadings to organize the information you collect from your literature search. Use the italicized headings ***Literature Source, Literature Procedure, Safety Precautions,*** **and** ***Stoichiometric Calculations.***

9. Under ***Literature Source***, simply copy the reference listing we give you on the sheet containing the identity of your compound. If you were doing a full literature search for an unknown reference, this section would have to be a lot longer and include all the various indexing references such as *Chemical Abstracts*, but here it'll be only a couple of lines.

10. Under ***Literature Procedure***, either copy out—word-for-word—the exact procedure given in your reference, or photocopy the relevant section and tape it into your notebook. If the procedure isn't very detailed, use the discussion later (in Part I: Synthesis, under Literature References) to turn the reference's vague procedure into a detailed one you can use. Write out this expanded version in addition to the reference quote, unless the reference is really detailed.

11. Under ***Safety Precautions***, copy down the safety data as described later (Part I: Synthesis, under Safety Precautions). This is an important precaution—chemicals won't hurt you if you respect them, but you have to know what "respect" means.

12. Under ***Stoichiometric Calculations***, work out the amounts you'll need for the size preparation you want to make. This set of calculations is also described later (Part I: Synthesis, under Literature References), but do them in this part of your notebook before you actually start the wet stuff in the lab.

13. Once you get to the actual synthesis and analysis of your compound, use the following italicized headings to organize the material in your notebook: ***Purpose, Procedure, Apparatus, Data, Calculated Results, Summary and Conclusions***. Each of these should appear for each separate experiment within a given project. In addition, at the end of the project as a whole, there should be a separate underlined heading ***Final Project Summary*** with its own discussion. These headings and the associated writeup are discussed further below. **EACH SHOULD BE ON A RIGHT-HAND PAGE!**

14. Under ***Purpose*** put about a one-sentence statement of the reasons for the experiment and the kind of information you hope to get by doing it. Keep it short, but think about the experiment's purpose and put it in your own words.

15. Under ***Procedure,*** outline exactly what you expect to do in that day's work. Be exact, but also remember it's only an outline and keep it reasonably short. It'll help if you put the procedure in the form of a numbered list of things to do. It'll also help us in grading it if you write on alternate lines instead of every line. Even done this way, there's no experiment so complicated that you can't outline a procedure in two-thirds of a page. The level of detail needs to be such that another general chem student with no background other than yours could pick up the notebook, read the procedure, and know what to do in order to repeat the work.

16. Under ***Apparatus***, list the individual pieces of equipment you'll need; no need to say how you're going to use them. Be sure to include the chemicals needed along with the pieces of hardware. IF (but only if) you're going to assemble the hardware into a complicated arrangement, you should sketch the arrangement you have in mind on the left-hand page opposite the apparatus list. But don't spend time drawing elaborate pictures of a beaker and a stirring rod! When you use a major instrument, note the manufacturer and model (for instance, "Digilab Excalibur infrared spectrophotometer").

17. Under ***Data***, there are two kinds of things you could get: a set of numbers (for example, when you weigh out samples for analysis you would need to record the weights) or simply a description of your visual observations of a process that's going on (for example, a chemical reaction mixture might change from yellow to dark green while it boils down for half an hour). If you're just describing a process that doesn't involve any important numbers, you'll only need to write the heading *Data*, nothing else in advance. On the other hand, if you expect to get a series of numbers you should set up a blank table so the numbers will appear in orderly, labeled columns.

18. Now stop and think a minute. You should have done steps 13-17 before coming into lab in the afternoon, and should have a fairly good idea of what you expect to do. If you're going to do the kind of thing that will obviously only require one try, regardless of the results, you can plan on doing the next day's work on the next right-hand page. On the other hand, if you're accumulating numerical data you may feel that you need to repeat the experiment several times—so if this is even a possibility, you should leave the next two or three right-hand pages blank. In other words, space the different experiments of a given project out through the notebook to some extent so that you can keep all the titrations together, all the spectra together, etc. This is the reason for the table of contents—it'll help you keep track of where you did certain kinds of work.

19. The heading ***Calculated Results*** may not need to appear. You'll need it if you're getting numbers that you have to treat in some way (such as calculating percent composition, for example). On the other hand, if you're not getting numbers or if you're getting numbers like melting points that don't require any calculations, you won't need this heading at all. If you do need it, set up the algebra and numbers for a sample calculation on the left-hand page opposite this heading: work that sample

calculation out in detail, but do all the other calculations like it on a calculator and just enter the numerical results in tabular form on the right-hand page under the heading. With units, of course.

20. After you've finished all of the above for a given experiment—doing the experimental setup, making the observations, running through the calculations—you should go on to the ***Summary and Conclusions*** heading. This should include averages and standard deviations of any calculated numerical data, discussion of the reliability and physical meaning of the numbers of other observations, and a brief indication of what this result means you ought to do next. Also, if you have any improvements on the overall procedure you ought to mention them.

21. At the end of an entire project, when you're getting ready to turn the notebook in, the ***Final Project Summary*** is particularly important. This should include summaries of your numerical data or calculations, comments on their meaning and reliability, an overall discussion of the project as a whole and a full account of your net result and the meaning you can attach to it.

22. Now, some general comments on notebook technique. First **YOU'VE GOT TO USE THE NOTEBOOK ITSELF FOR ALL—REPEAT *ALL*—OF THE PAPERWORK YOU DO IN THE LAB.** Don't write things down on other pieces of paper; use the notebook as the *only paper you ever write on* for Chem 150. This is true even if you think what you're writing down may be wrong. If it is wrong and you learn from it and then do it right, we won't count it against you. Furthermore, if we see you writing on other non-notebook paper, we'll take it and keep it, so that you'll have the work to do over again. So don't do it!! And remember that we will expect you to pledge that you have recorded data in your notebook immediately, not when you get back to your room!

23. This brings us to the next topic, which is how to treat mistakes. (They'll be fairly frequent in your notebook—they are in everybody's.) There are two kinds of mistakes. One is a minor misreading of a number from a balance or some other instrument; the other is the sudden realization that everything you've done for the last two hours is screwed up, so that you have a whole page or even three or four pages wrong. For the minor misreading, simply *draw a single line through the mistake* and write the correct number or word above it; don't bother explaining it. But for a major goof of several lines or a page, *draw a big X through the area to be disregarded* and put a few words of explanation opposite the goof (left-hand page). It won't count against you at all, and you can go on from there. The reason for prescribing this format for corrections is that occasionally "mistakes" turn out to be useful after all, and they need to be readable even through the correction. So: **Mistakes don't count against you** (at least not recognized ones), but improper corrections do cost you a little bit.

24. There are some other minor points of notebook keeping that we insist on. You must record the units with any number you put down. And the pledge in its proper form (see page 8) must immediately follow the final project summary in the notebook. Violating any of these ground rules will cost you a few points.

25. At several points through the manual there will be a double-page spread that will give you an idea how the notebook ought to look after you get into that part of the project. Note that numerical data is in tabular form, and that it's spaced out so as to be reasonably easy to read. Observations are recorded in a list. The right-hand page has the headings mentioned earlier, and the left-hand page has calculations.

26. One last note: the lab lectures have to stay a week or so *ahead* of you, so you need to take written lab lecture notes (upside down in the back of your notebook) so you can go back to them for help. Each lecture needs a title and about 3/4 page of notes, with details on things to do and not to do.

CHECKIN INSTRUCTIONS

During the first laboratory period you will be assigned a laboratory desk containing an assortment of standard equipment. Since a good bit of that equipment is glassware that must be replaced if broken, we require a breakage deposit of $35.00 paid to the stockroom manager before checking in. Besides the cost of broken equipment, the cost of this lab manual ($7.00) and of the lab notebook ($6.00) you will use are deducted from the deposit before refunding it to you at the end of the semester. That is, a man with a steady hand can get $22.00 back. If you're going on to Ch152 in the spring, we'll give you another notebook and charge you another $7.00 then.

This manual and the lab notebook will be handed out at a lab lecture in the classroom on the first scheduled day of lab. We'll also give you a desk inventory sheet like the one on the opposite page, with your desk number filled in. When we get to the end of the explanation of checkin, follow the procedure below.

1. Find your desk in the laboratory and open the combination lock on the desk. The lock should be open so that you don't need the combination. Write your desk number down somewhere so you can remember it next week!

2. At some point in the next 30 minutes, go to the stockroom and pay your breakage deposit. When you pay it, the stockroom manager will give you your lock combination.

3. Write this combination down on something you always have with you or have it tattooed on your arm. DON'T write it on your lab manual or on the front of the lab desk!

4. Remove the equipment from your desk, be sure it's clean and not chipped or cracked, and compare it with the desk inventory form. Be sure you can identify each item; pictures are on the next few pages.

5. If any items are missing, cracked, or broken, list them on the desk inventory sheet and get them from the stockroom manager. DON'T GO IN THE STOCKROOM.

6. When you have a complete set of equipment and recognize all of the pieces, sign your desk inventory sheet with the items checked. Note that there's a copy of the desk inventory sheet on the opposite page for your reference.

7. Put everything back in your drawers or cabinet and **LOCK YOUR DESK.**

8. Turn the signed desk inventory sheet in at the stockroom. **FROM THIS POINT ON YOU ARE RESPONSIBLE FOR THE CONTENTS OF YOUR DESK.**

You'll need to start keeping a table of contents for your notebook as soon as you start on it. We've reproduced the first page of Irv Gratch's table on the next page to help you get started. Incidentally, all his notebook page dates are historically significant–can you decipher them?

1

Irv Gratch
8/24/14

TABLE OF CONTENTS

Expt 1	Literature Search for Synthesis of $ZnCl_2 \cdot 2\,C_5H_5N$	5
	Literature Source for Synthesis	5
	Literature Procedure	5
	Safety Precautions	7
	Stoichiometric Calculations	8
Expt 2	Synthesis of $ZnCl_2 \cdot 2\,C_5H_5N$	11
	Purpose	11
	Procedure	11
	Apparatus	11
	Data	13
	Calculated Results	13
	Summary + Conclusions	13
Expt 3	EDTA and Zn Solution Prep	15
	Purpose	15
	Procedure	15
	Apparatus	15
	Data	17
	Calculated Results	17
	Summary + Conclusions	17
Expt 4	EDTA Standardization Against Known Zn Solution	19
	Purpose	19
	Procedure	19

cont.

COMMON CHEMICAL APPARATUS

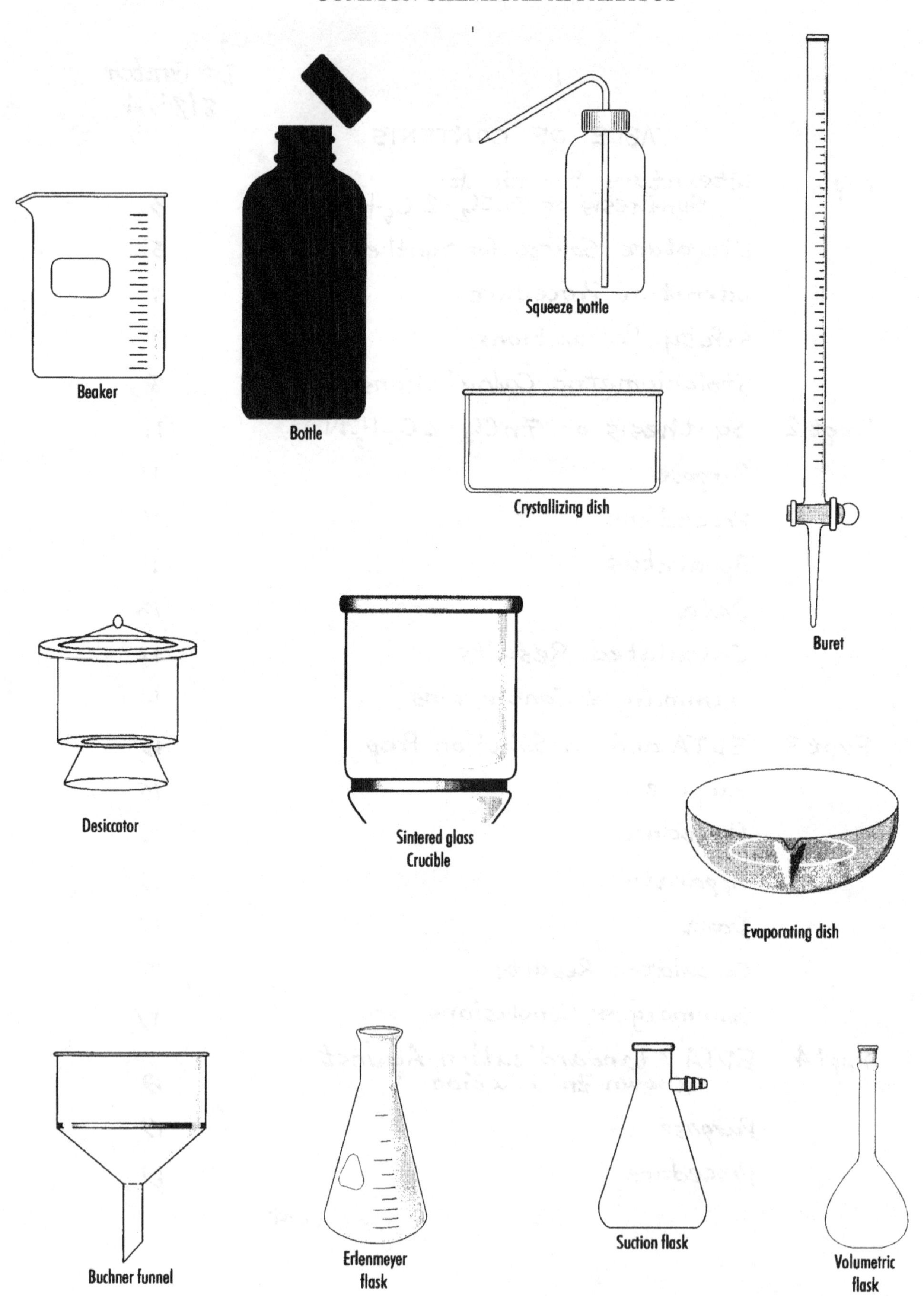

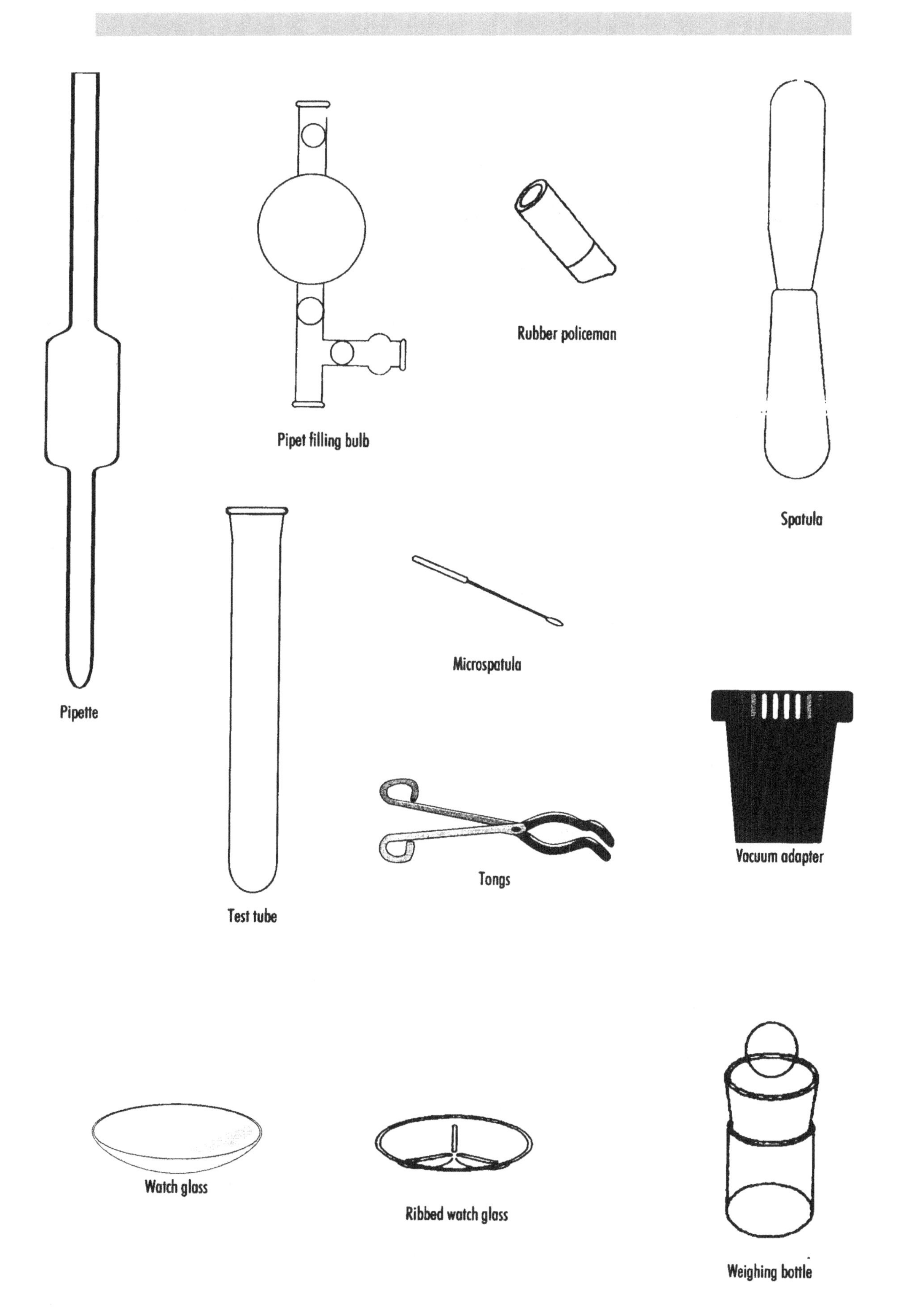
Pipet filling bulb
Rubber policeman
Spatula
Pipette
Microspatula
Test tube
Tongs
Vacuum adapter
Watch glass
Ribbed watch glass
Weighing bottle

FALL SEMESTER PROJECT: COORDINATION COMPOUND SYNTHESIS AND ANALYSIS

In this laboratory, you will each be working on your own unique project for the whole semester. Yours will be at least a bit different from all the others, although they will all be generally similar. In your project, you will be given a sheet of paper with the formula and official name of a ***transition metal coordination compound*** (or ***metal complex***), and with a library reference to a book or journal article that describes its preparation. Each student will have a different coordination compound, so the syntheses will be different.

What you should do is to look up the synthesis by going on-line to the H-SC library (we'll show you how to do that), consult with the lab instructor on your approach, prepare 10 grams of the coordination compound, and then analyze it for its metal content and halide content (all assigned compounds contain a metal atom and either bromide or chloride ion). In the analysis, you will first familiarize yourself with the analytical procedure by "running a known"—performing the same analysis on a cheap, plentiful compound with a known composition that you expect to run on your coordination compound. So there will be four analyses: two halide analyses and two metal analyses. These will be done by what are called ***wet methods***, in which quantitative chemical reactions are run in solution to determine the amount of material present. Then you will perform a ***spectroscopic*** analysis on your compound for its metal content to compare with the value obtained in the wet analysis. For the metal and halide analyses, you will report the weight percent of the element present in your compound and compare your values with the theoretical values for that formula.

PART I: SYNTHESIS

In this part of the project, you'll be making ten grams of the coordination compound that is on the sheet issued in the lab lecture. Although this is obviously laboratory work, you won't be working in the laboratory at first; instead, you'll be doing what chemists always do, which is to check the literature background for the work at hand. However, although the work is taking place in the library instead of the lab, it's still laboratory preparation, and all of your library work should be described in your notebook! So the first thing to do is to make sure your notebook is prepared in the fashion described on pages 30-33 of this manual. Use the on-line journal PDF or look up the book in the library, and take notes on the procedure with references in the following format:

For a reference from a book (published one time):

Author last name, initials, title underlined, edition (if any), city of publication, publisher, year of publication, pages referred to.

Example: Porterfield, W. W., Inorganic Chemistry, Reading, MA, Addison-Wesley, 1984, pp. 487-494.

For a reference from a journal:

Author last name, initials, abbreviated journal name, year double underlined, volume underlined once, initial page number of the article.

Example: Axtell, D. D. et al., J. Am. Chem. Soc. 1973, 95, 4555.

Note that if there are two authors both should be named, but if there are more than two list only the first, followed by et al.

You'll need a list of the official abbreviations for journals; these are the ones in our library that you might wind up using:

Journal of the American Chemical Society	**J. Am. Chem. Soc.**
Inorganic Chemistry	**Inorg. Chem.**
Journal of the Chemical Society	**J. Chem. Soc.**
Journal of Inorganic and Nuclear Chemistry	**J. Inorg. Nucl. Chem.**

We have the first two on-line; for the second two you'll need to email your instructor for the PDF. Sometimes literature references will give you a procedure in good detail, so that you can follow it simply by scaling the amounts to give you a ten-gram yield. Alas, most of the time they only have a very skimpy procedure, because they think the chemist reading the procedure knows all the usual basic stuff. So when you find the procedure, copy it down in detail (maybe Xerox it and mount the photocopy in your notebook), then consult the lab instructor for advice on amplifying a skimpy procedure into something you can follow in detail.

TRANSLATING LITERATURE PROCEDURES

Your synthetic procedure will either be from a research journal article, or from the periodical volumes *Inorganic Syntheses*. Because journal-article authors think other research chemists are reading the material, they can be pretty vague sometimes; on the other hand, *Inorganic Syntheses* preps are magnificently detailed (though of course they aren't generally going to be set up for exactly ten grams of your compound). In the next couple of pages we'll provide some help with "translating" these two types of reference into language you can work from.

Suppose your coordination compound is "**Bis(pyridine)dichlorozinc, $ZnCl_2 \cdot 2NC_5H_5$**". Then in general you would expect to form it by reacting the ligand pyridine with the metal salt zinc chloride as in the reaction equation below. Let's look at how this prep might appear in, first, a journal article, and then as it might appear in *Inorganic Syntheses*. Remember that the general process for all these coordination compounds is going to look like this:

metal halide + ligand → coordination cpd

$$ZnCl_2 \cdot 2H_2O \quad + \quad 2\,NC_5H_5 \quad \rightarrow \quad ZnCl_2 \cdot 2NC_5H_5 \quad + \quad 2\,H_2O$$

1 *zinc chloride*   2 *pyridines*   1 *bis(pyridine)dichlorozinc*   2 *waters*

Journal-type Procedure
"Appropriate mole ratios of the anhydrous metal salt and ligand were mixed in hot ethanolic solution; crystals formed on cooling and were dried under vacuum."

Translation:

The "metal salt" is zinc chloride, with no waters on it ("anhydrous"). The "ligand" is pyridine, which is a liquid. "Appropriate mole ratios" means that mole amounts should be used in the ratio that they appear in the balanced reaction equation, which means 2 moles of pyridine for every mole of zinc chloride. A mole is one formula weight quantity, so we need to add up the atomic weights to get the formula weight of zinc chloride (65.37 + 2x35.45 = 136.27 g $ZnCl_2$/mole) and of pyridine (14.01 + 5x12.01 + 5x1.008 = 79.10 g NC_5H_5/mole). The 2:1 mole calculations are below.

When we get the right weights to use (4.63 g $ZnCl_2$ and 5.37 g pyridine—from the calculations below), each of the two substances should be weighed out into a small beaker so they can be dissolved. Usually we'll need about 5 mL of solvent for each gram of stuff to be dissolved, so in this case each chemical will need about 25 mL of solvent. The solvent is ethanol (ethyl alcohol); ethanol comes two ways, called 95% (meaning 5% water) and 100% or "absolute ethanol". If your procedure calls for "absolute" or "anhydrous" ethanol, be sure to use that—but if it doesn't specify either type, use 95%, which is a lot cheaper.

When both substances have been dissolved, each in its own beaker, we'll slowly add the contents of one beaker to the other beaker to let the reaction occur. This means that at least one of the beakers has to be big enough to hold the final total volume, so think about that before you choose beakers to weigh into. The procedure specifies "hot ethanolic solution", so the beakers should be heated on a hot plate in a fume hood at the side of the lab before mixing.

After the two solutions have been combined, solid crystals of product may form immediately—but the procedure suggests that the solution will have to be chilled before solid separates from the solution. Plan to use a pan of crushed ice from the ice machine in the hall, with about a half inch of water in it to improve thermal contact with the beaker. If this chilling doesn't yield anything, we can put the solution into the laboratory freezer overnight—but consult before you do that.

THE STOICHIOMETRIC WEIGHT CALCULATION FOR A JOURNAL-TYPE PREP

Start with a target weight of 10 g product and let the units work it for you:

$$? \ g \ ZnCl_2 = 10 \ g \ ZnCl_2{\cdot}2C_5H_5N \times \frac{1 \ mol \ ZnCl_2{\cdot}2C_5H_5N}{294.48 \ g \ ZnCl_2{\cdot}2C_5H_5N} \times \frac{1 \ mol \ ZnCl_2}{1 \ mol \ ZnCl_2{\cdot}2C_5H_5N} \times \frac{136.27 \ g \ ZnCl_2}{1 \ mol \ ZnCl_2}$$

$$= 4.63 \ g \ ZnCl_2$$

Calculate for the ligand weight in the same way, but remember to use 2 moles!

<u>Inorganic Syntheses</u> Procedure
"12.48 g $ZnCl_2$ was dissolved in 75 mL 95% ethanol at 60°C and added slowly to a solution of 28.65 g pyridine in 150 mL 95% ethanol at 60°C. On slow cooling to 0°C in an ice bath, white crystals formed, which were filtered using a Buchner funnel and dried for 12 hours in a vacuum desiccator. The yield was 39.65 g (94%)."

Translation:

This doesn't need nearly as much translation. Use a beaker with a volume 2 to 4 times that of the solution you expect to make. Be sure *all* of solid A is dissolved before component B is added. If heating is necessary, do it in a fume hood on a hot plate.

The "Buchner funnel" business refers to the process of separating the solid product from the liquid solution in which it formed by filtration. Filtration and drying techniques are discussed on page 46.

A YIELD-SCALING CALCULATION FOR AN INORGANIC SYNTHESES *PREP*

In this kind of literature reference, the detailed instructions tell you how much the original preparation gave (which might not have been 100% of what would be expected). Here it's particularly simple to work out the amounts to use for a 10-gram preparation. Just scale the amounts used in the literature prep by a factor that's equal to your yield (10 g) divided by their yield (here 39.65 g):

$$\begin{aligned} ?\ g\ ZnCl_2 &= 12.48\ g\ ZnCl_2 \times \frac{10\ g\ ZnCl_2{\cdot}2C_5H_5N}{39.65\ g\ ZnCl_2{\cdot}2C_5H_5N} \\ &= 3.15\ g\ ZnCl_2 \end{aligned}$$

Do the same kind of calculation for the ligand and for the volumes of solvents that will be used.

SAFETY PRECAUTIONS

After you've looked up your literature reference for the synthesis and worked out amounts of each reactant to use, it's important to assess how much of a chemical hazard these chemicals are. We're not going to assign anything that's a potential nerve gas, of course, but even common solvents can sometimes have some risk in terms of skin contact, etc., so we want you to write up in your notebook a short assessment of the health hazard associated with each chemical you're going to use.

Once you have the specific chemicals identified by name and formula, look each of them up in *The Sigma-Aldrich Library of Chemical Safety Data*, which is a reference book (in two volumes) that we keep in Room 220. It's alphabetical and in tabular format. For each chemical or solvent you're going to use, copy down what is shown under the heading *Health Hazards*, and read the rather drastic instructions under *Handling and Storage*. Ask us about how much of the handling precautions you need to follow (face shield, gloves, etc.). This is an important precaution, and we won't approve your synthesis procedure until you have this in your notebook! One additional note: You can also find useful toxic-hazard information in the *Merck Index*, which is in the library and in several faculty offices.

COMMON TECHNIQUES

We might note at the beginning that there are two good short books on lab technique—how to do what you want to do—in the departmental book collection in room 220: *Laboratory Technique in Organic Chemistry*, by Wiberg, and *Principles and Techniques for an Integrated Chemistry Laboratory*, by Aikens *et al.* These both have concise and very helpful discussions of nearly all the techniques you'll need. However, we can provide a quick summary here of a few techniques that you'll almost certainly run into, even though the projects do differ.

Reactions between solids and liquids or between two liquids are usually carried out in a ***beaker*** or a ***flask***. Beakers allow more convenient removal of solids; flasks are more easily stoppered and prevent splashing better. Reactions between two solids are rare but are usually carried out in a beaker or sometimes in an ***evaporating dish*** or a ***crucible***. Reactions between a liquid and a gas are usually carried out in a flask unless removal of a solid will be important, in which case a beaker is usually used. If your reaction is to be carried out by dissolving a solid in a liquid solvent (usually the case), but there's no specific advice on relative amounts of solute and solvent, assume 1 gram of solid will dissolve in 5 mL of liquid. That's a very rough guide, but it's usually in the right vicinity.

If liquids are to be added slowly to a reaction mixture, an ***addition funnel*** is usually used; this is essentially a big funnel with a stopcock in the stem that may be opened very slightly to allow slow passage of the liquid in the body of the funnel. If gases are to be passed into liquids, the addition is usually performed with a ***gas bubbler,*** a porous glass disk at the end of an immersed tube that provides many small bubbles and thus good gas-liquid contact. Stirring is usually provided for any liquid-phase reaction to assure good mixing of the reactants, usually just using a stirring rod. However, prolonged stirring will usually mean using a ***magnetic stirrer*** (a little plastic-coated bar magnet in the beaker, driven by an electric motor and magnet underneath).

Heating will always be done electrically in this lab, using a ***hot plate***, a stirring hot plate (with a magnetic stirrer in it), or a ***heating mantle***. The mantle is a flexible heated hemisphere, like a small fat electric blanket, into which round-bottomed flasks fit snugly; it provides very even heat. Cooling may be done with an ice bath (to 0°C), an ice-salt bath (to -20°C), a dry ice-acetone bath in a ***Dewar flask*** (like a thermos bottle; to -75°C), or liquid nitrogen in a Dewar flask (-195°C). Heating or any other process that involves the possible generation of vapor or other gases should be done in the fume hood.

If at the end of the reaction you need to get crystals to come out of solution, and even if they *have* crystallized out, you should chill the solution (the ***supernatant*** liquid) to improve the yield; crystals are almost always less soluble in a cold liquid. You can use an ice bath for this, but if crystals won't form try covering the beaker with Parafilm (like fat Saran wrap) from the stockroom and placing it in the freezer overnight. (But don't put water in the freezer!) Another trick for knocking crystals out of solution is to change the polarity of the solvent by adding a small quantity of another solvent to it. The lab instructor will help you with this if you have trouble getting crystallization to occur.

WRITING UP A SYNTHESIS PROCEDURE

In your notebook at this stage you should have finished up the library reference, done the stoichiometric calculations to work out amounts of each chemical to use, and copied the safety data (the Aldrich *Health Hazards*). It's time to make the stuff!

In the general notebook instructions on page 30-32 you should have followed steps 1-6 before last week and then steps 7-12 during last week's lab. Now it's time to start using the notebook in the lab, and there's a standard (and easy to use) format for writing up the notebook, both in planning work ahead of time and when you're actually in the lab watching things happen. The basic structure of the notebook for each experiment is itemized in step 13 on page 31:

- **Purpose (step 14)**
- **Procedure (step 15)**
- **Apparatus (step 16)**
- **Data (step 17 — include observations of color changes, etc., and weight yield)**
- **Calculated Results (step 19 — percent yield)**
- **Summary and Conclusions (step 20)**

First (step 14) you need a *Purpose:* If you don't know where you want to go, any road will take you there.

Then (step 15) you need a detailed, list-type *Procedure* that you write up ahead of time, outlining in detail the sequence of actions you expect to take. In fact, you'll wind up doing at least a few things in a different way—but that's what the later section of the notebook called *Data* is for. What follows here is what would be a pretty good procedure for making the zinc-chloride/pyridine complex that has already been described above, $ZnCl_2 \cdot 2NC_5H_5$. Pages 44-45 show typical left-hand and right-hand notebook pages for this step.

Procedure: ***[Don't copy this procedure – it's just an example of the format and information!]***

1) Weigh 4.63 g $ZnCl_2 \cdot 2H_2O$ into 150-mL beaker on toploading balance.

2) Weigh 5.37 g liquid pyridine into 100-mL beaker on toploading balance.

3) Add 25 mL 95% ethanol to $ZnCl_2$ in beaker and stir to dissolve completely.

4) Add 30 mL 95% ethanol to pyridine in beaker and stir to dissolve completely.

5) Heat both beakers on hot plate in hood until first signs of boiling occur.

6) Remove beakers from hot plate and immediately add contents of pyridine beaker to $ZnCl_2$ beaker while both solutions are still hot.

7) Allow beaker with reaction mixture to cool 5 min covered by watch glass, then place covered beaker in crushed-ice bath for 30 min to form solid product.

8) Set up suction filtration apparatus as in sketch on next page, with suction flask, rubber vacuum adapter, Buchner funnel, and filter paper.

9) Filter reaction mixture through Buchner funnel, using rubber dam to press solid product dry.

10) Transfer damp product to porcelain evaporating dish and dry in vacuum desiccator for 60 min.

11) Weigh empty storage bottle with cap.

12) Transfer dried solid to storage bottle, cap tightly, and weigh to obtain yield.

13) Store labeled bottle in desiccator.

Again, the next two pages show the way notebook pages ought to look in the synthesis part of the project.

One way or another, you're likely to reach a stage at which you need to separate solid crystals of your compound from a liquid solution. This means filtering the mixture; for most of you, the easy way to do this is to use a ***Buchner funnel*** and suction filtration. You'll need the water ***aspirator*** on the sink (the little side-arm on the faucet exit), heavy-wall rubber tubing, your ***suction flask***, the ***vacuum filtration adapter***, the Buchner funnel with a piece of ***filter paper*** the right size to fit it lying flat (covering all the holes but not folding over at the sides), and a piece of ***rubber dam*** (thin latex sheet) with a rubber band.

8

$py = C_5H_5N$

$$? \text{ g } py = 10 \text{ g } ZnCl_2 \cdot 2py \times \frac{1 \text{ mol } ZnCl_2 \cdot 2py}{294.18 \text{ g } ZnCl_2 \cdot 2py} \times \frac{2 \text{ mol } py}{1 \text{ mol } ZnCl_2 \cdot 2py} \times \frac{79.10 \text{ g } py}{1 \text{ mol } py}$$

$= 5.372 \text{ g } C_5H_5N$ target wt.

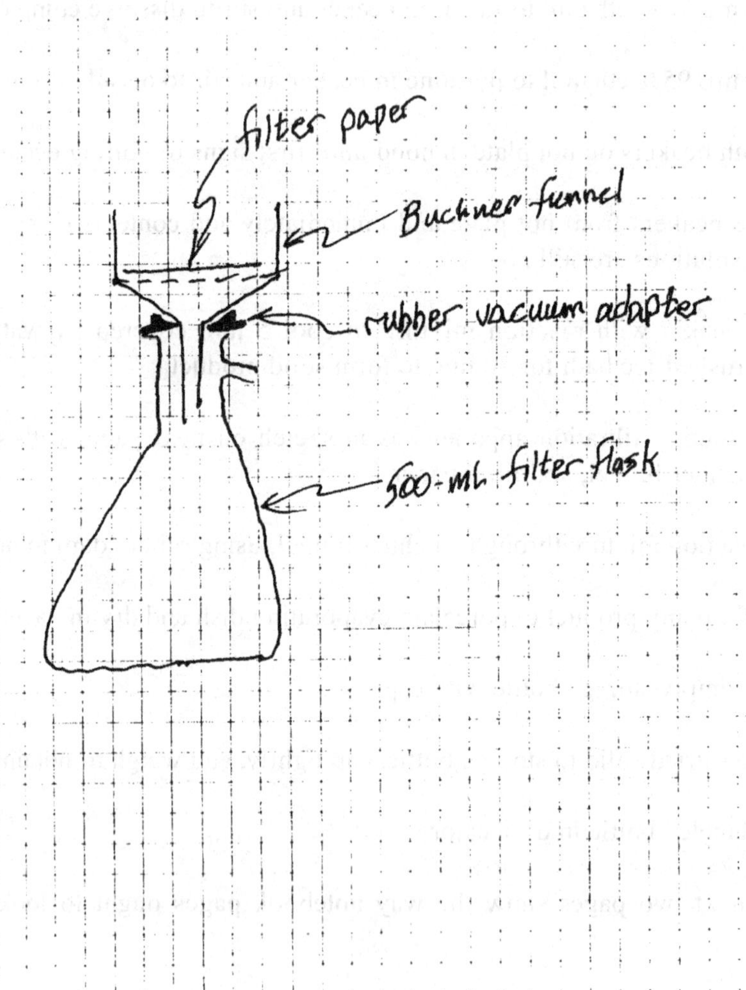

9

Irv Gratch
9-1-39

EXPERIMENT 2: Coordination Compound Synthesis

Purpose: To prepare 10.0 g $ZnCl_2 \cdot 2C_5H_5N$ for analysis

Procedure:

1) Weigh 4.63 g anhydrous $ZnCl_2$ into 150-mL beaker on toploading balance
2) Weigh 5.37 g pyridine into 100-mL beaker on toploading balance.
3) Add 25 mL 95% ethanol to $ZnCl_2$, stir to dissolve completely.
4) Add 30 mL 95% ethanol to pyridine, stir.
5) Heat both beakers (hot plate in hood) till first signs of boiling.
6) Remove beakers from hot plate + combine immediately.
7) Allow beaker with reaction mixture to cool 5 min, then chill 30 min in ice bath.
8) Set up suction filtration apparatus (at left).
9) Filter cold reaction mixture through Buchner funnel; press dry with rubber dam.
10) Transfer solid product to evaporating dish, dry 60 min in vacuum desiccator.
11) Weigh empty storage bottle with cap.
12) Transfer dry product to bottle, weigh capped bottle.
13) Calculate wt. yield + % yield.

Apparatus: 100-mL beaker, 150-mL beaker, toploading balance, 2 stirring rods, hot plate, ice bath, filtration flask, Buchner funnel, vacuum adapter, rubber dam, vacuum desiccator, SC bottle, filter paper

Data:

1) $ZnCl_2$ wt. 4.64 g, pyridine wt. 5.35 g
2) $ZnCl_2$ solution colorless, pyridine solution light yellow.

Set the filtration apparatus up as the sketch at the right suggests. The filtration adapter should have the conical flange on top if it's flat; if the whole rubber adapter is conical do it the obvious way. Turn the water on full force before hooking the tubing up to the flask and make sure that you're getting a vacuum. Then attach the tubing, press down on the funnel, and pour a little of the pure solvent that you're using onto the paper in the bottom of the funnel to seal it to the funnel (like the wet T-shirt contest).

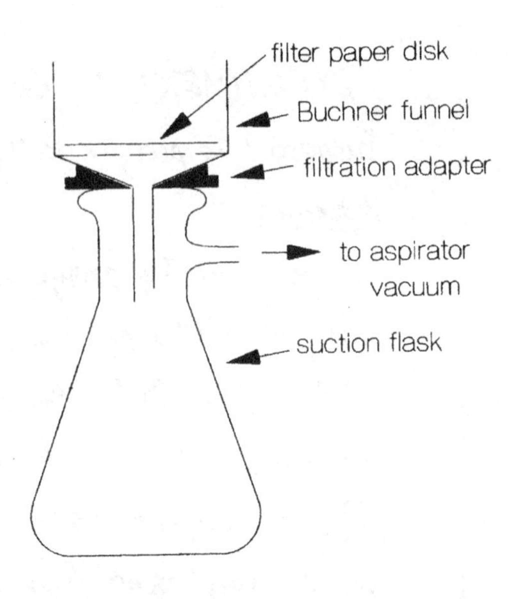

Pour the solid-liquid mixture into the funnel, using a *little* solvent (2 or 3 mL) to rinse the last crystals out of the beaker. Once the last liquid is in the funnel, place the rubber dam over the top of the funnel and use a rubber band to seal it to the funnel. When the last liquid sucks through, atmospheric pressure will force the rubber dam down against the crystals to squeeze the liquid out. Let it squeeze the crystals dry for about a minute, then remove the tubing from the sidearm on the suction flask.

Turn off the faucet. Scrape the crystals from the underside of the rubber dam into your porcelain evaporating dish using your spatula. Use the spatula to remove the mat of crystals from the funnel, transferring it to the evaporating dish. Remove the filter paper, and chop the mat of crystals up with your spatula to increase their surface area. Ask an instructor to put your crystals in a vacuum desiccator to strip off the last bit of solvent—if the solvent is water this will take at least 90 minutes, while ethanol takes about 45 minutes and ether 30 minutes.

While the crystals are drying, label your small brown screw-cap bottle with your name, the desired compound's formula, and the date, leaving space enough on the label for the weight of the compound. Use the toploader to get a ***tare weight*** (empty weight) of the bottle with cap and record that in your notebook. When the crystals are dry, transfer them to the bottle, reweigh it to get the yield, and write the weight of crystals on the label. And, of course, in your notebook. **Store the bottle, tightly capped, in your desiccator.**

You'll have leftover chemicals: metal salts, solvents, maybe some organic ligand. Proper chemical waste disposal is important! Ethanol (ethyl alcohol) can be poured into the sink with the water running. Other chemicals should be placed in containers labeled **"HAZARDOUS WASTE"** in the lab; in general, each container will also be labeled with the type of waste to be placed in it (for instance, **CHLOROFORM** or **METAL SALTS**). ***Depending on the contents, one bottle can cost the College up to several hundred dollars to dispose of, so proper identification is critically important. Do it right!***

CALCULATIONS AND PROJECT SUMMARY

Use the weight of crystals to calculate a percent yield for your reaction. This means working out which is the limiting reagent, calculating the theoretical yield on that basis, and expressing your experimental yield as a percentage of that. Don't use the literature yield as the theoretical yield—do the limiting-reagent calculation.

Your notebook discussion, at this point, should contain all of the headings for the synthesis described in item 13 on page 30; the ***Data*** in this case will be a description of the actual weights and volumes you used plus your visual observation of what happened when you ran the reaction (see item 17, p. 30, and the sample notebook pages on pp. 42-43). The ***Calculated Results*** will contain your percent yield (with actual calculations or calculator tape on the left-hand page).

Your notebook is due for a first-stage grading at the end of the fourth lab period. You should have finished the synthesis, but *the notebook is due regardless* and should be up to date.

PART II: ANALYSIS

With a bottle of your coordination compound (you hope) in hand, it's time to analyze it to find out whether you really did make what you set out to make. All of the compounds should contain a transition metal (Mn, Fe, Co, Ni, Cu, Zn, or Cd), a halide (here only Br or Cl), and almost always an organic species as a ***ligand*** (electron pair donor) or as a ***counterion*** (positive ion for a negatively-charged complex ion). We can analyze easily for the metal and for the halide, and can get evidence of the presence of the organic part.

ANALYTICAL MEASUREMENT ACCURACY

But first some thoughts about quantitative analysis! In the synthesis you've just run, none of the weights or volumes needed to be any more accurate than about 1% error; sometimes 10% slop was acceptable. This is because the ten-gram prep you wanted could really be 9.7 g or 10.3 g without its making any difference. So, for example, you used the top-loading balance to get 6.24 g of a reactant—plenty accurate enough. In analysis, however, the rules are different! You should have noted in your lecture text that the stoichiometry problems usually have data good to about one part per thousand (94.2% is good to one little number in 942 little numbers). Because we can measure weights and volumes to one part per thousand, you should be certain to achieve that accuracy (one part per thousand or, usually, four sig figs) in your analysis! Using the Sartorius analytical balances, for example, the intrinsic error is 0.0002 g—so to achieve one part per thousand (1 ppt) error we only have to take sample weights of 0.2000 g or thereabouts. And when you express your final weight percent metal or halide, you should take the value to two decimals: NaI is 84.66% I, for example. That's 8466 little numbers, so 1 ppt precision would require a reproducibility of about 8 little numbers, or 0.08 standard deviation for a series of measurements. So it's necessary to work very carefully, very neatly in quantitative analysis; watch for even the least trace of dirt or crumb of material to be weighed.

One of the things this means is that you'll need to use deionized water, not tap water, whenever you make up solutions in quantitative analysis. In particular, tap water has significant amounts of chloride in it that would certainly mess up your halide analysis. The deionized water faucet is at the sink on the side wall of the lab. Don't waste it (rinse with small quantities!) but *do* use it.

In order to establish reproducibility for your analyses, you'll need at least triplicate results that, as far as you know, don't have anything wrong with them. So when we tell you to weigh out a sample, weigh out three samples in three beakers or flasks, etc. Generally, running triplicate samples only takes a little longer than running a single sample because they can all wait for something to happen at the same time: dissolve, crystallize, etc.

VOLUMETRIC METAL ANALYSIS

In the metal analysis you will perform a quantitative chemical reaction that ties all of your metal up as a water-soluble complex, and you'll measure the volume of a reacting solution of known molar concentration to find out how many moles of metal you have.

In the general quantitative analysis sense, there are two new concerns about accuracy: (1) How do you measure volume with 1 ppt accuracy? (2) How do you get a reacting solution with known concentration to 1 ppt accuracy? The answer to question 1 is that you'll use a buret, which measures up to 50 mL to the nearest 0.02 mL with about 0.05 mL error (or 1 ppt). You'll have to read the buret to the nearest 0.02 mL; this will mean estimating even numbers (0.02, 0.04, etc.) between the finest graduations on the buret, which are 0.1 mL apart. Always read the buret to two decimal places! Estimate by using your buret reader with the black/white boundary slightly below the bottom of the liquid surface (the ***meniscus***) in the buret, and judge the location of the bottom of the meniscus.

Answering question 2 requires us to know something about the nature of the quantitative chemical reaction that's going on. The species in solution that reacts with the metal ion is the ***ethylenediaminetetraacetate*** ion (often abbreviated Y^{4-}), which comes from ethylenediaminetetraacetic acid (H_4Y or EDTA). The EDTA ion can wrap around the metal ion and tie it up in a very stable, quantitative fashion.

So the chemical reaction (regardless of which metal ion is involved) is:

$$M^{2+} + Y^{4-} \rightarrow MY^{2-}$$

which has 1:1 molar stoichiometry. Now all we need is an EDTA solution with its molar concentration known to 1 ppt.

The tetrasodium salt of EDTA, Na_4Y, in the lab is only about 98% pure, so we can't make up a 1-ppt-accuracy solution directly using the weight of Na_4Y. However, we can make up a solution that's approximately the right concentration and ***standardize*** it by allowing it to react with a known amount of a standard metal ion—zinc is the easy one to use. So there are two solutions to make up before starting on the volumetric metal determination: an EDTA solution that's approximately 0.02 M, and a zinc-ion solution that's also about 0.02 M. The EDTA (Na_4Y) solution can be made up in an approximate fashion, because it has to be standardized anyway—but the zinc solution has to be made up in **the most careful possible manner** in order to achieve 1 ppt accuracy in its concentration as prepared.

In your notebook, this is a new experiment. Start a new right-hand page and title it "EDTA Solution Preparation". Typical notebook pages follow on pp. 50-51 using the standard headings:

- **Purpose (step 14, page 30)**
- **Procedure (step 15)**
- **Apparatus (step 16)**
- **Data (step 17 – weights, volumes, any changes in procedure)**
- **Calculated Results (step 19 — molar concentration of Zn solution)**
- **Summary and Conclusions (step 20)**

Step one is to make up the EDTA solution. Look at the label on the bottle of "Ethylene-diaminetetraacetic acid, tetrasodium salt" in the lab to see what its molecular weight (or ***formula weight, F.W.***) is. Calculate the weight needed to make up one liter of 0.02 M solution. Weigh out that amount to the nearest 0.01 g on the top-loading balance. Dissolve it in about 600 mL of distilled water by stirring, transfer it to your clean 1-L plastic screw-cap bottle, and fill the bottle to the shoulder with distilled water. Cap it tightly and invert it 20 times to be sure the solution is thoroughly mixed. Label it with your name, the date, and leave space enough on the label for the concentration once you determine it accurately.

Step two is to make up the zinc solution, and this one has to be careful. Get a 1-cm piece of zinc wire from the lab desk, and make up the solution as follows: Wipe any fingerprints off the zinc wire with a Kimwipe, and weigh the wire on the analytical balance—to the nearest 0.0001 g—handling it with tweezers or a Kimwipe to keep more fingerprints from accumulating (you can actually weigh them). Place the weighed zinc metal (around 0.65 g or 0.01 mole Zn) in a clean 600-mL or 800-mL beaker, add 10 mL deionized water (use a graduated cylinder), place the beaker in a fume hood, and add 10 mL concentrated nitric acid (use a graduated cylinder again). Swirl gently until the zinc wire is completely dissolved. Make sure your 500-mL volumetric flask is clean and wet inside *only* with deionized water (at least three rinses of about 10 mL each). Pour the zinc solution into the volumetric flask. Rinse the beaker used for dissolving with four 50-mL portions of deionized water, adding each to the volumetric flask. Finally, dilute to the mark (match the **bottom** of the curved liquid surface—the meniscus—to the mark) on the flask with deionized water; look back at page 15 for a general discussion of volumetric flask use.

Once you have the zinc solution made up in the volumetric flask, invert it 20 times to mix it thoroughly, then use several small portions (about 5 mL) *of the zinc solution* to rinse out the glass screw-cap bottle, wetting the sides completely. Then transfer the rest of the zinc solution to the bottle and rinse out the flask. Calculate the molar concentration of the zinc solution (about 0.02 M) by converting the weight of the zinc wire to moles of zinc and dividing by the volume of 0.500 L. Label the bottle with your name, the date, and the molar concentration of the zinc solution **to 4 sig figs**: 0.01873 M, not 0.0187 M (remember, zero between the decimal and the first nonzero numeral *isn't* a sig fig).

Now you have an accurately-known zinc solution and an approximately-known EDTA solution, and you can get an accurate EDTA concentration by allowing them to react quantitatively with each other. This is a ***titration***, and we'll describe it below. Before you get into it, however, be sure that your 25-mL pipet and your buret are clean, in the sense that when they're filled with water, and the water is allowed to run out, the water remaining on the sides of the glass is in a uniform thin film, not in isolated droplets. Your buret should have been stored full of water, which will keep it clean, but your pipet probably has people grease inside it and will need to be cleaned. Use the "**Chromic Acid Cleaning Solution**" in the hood: fill the pipet above the line on the upper neck using a pipet filler bulb, and allow it to stand in the cleaning solution bottle for 5 minutes; then drain the cleaning solution back into the stock bottle, and rinse the pipet several times with tap water. The chromic acid solution is sodium dichromate in concentrated sulfuric acid, which is bad stuff—DON'T get it on you, and if you do get some on your skin or clothes, wash it off **immediately** and call the instructor!

When you store your buret or pipet, they will be wet; in fact, the buret should be stored full of water and corked. When you use either piece of glassware, rinse it out with two or three small (3-5 mL) portions of the *same* solution whose volume you're going to measure; then, and only then, fill it with that solution for measurement.

16

$? \text{ g } Na_4Y = 1 \text{ L soln} \times \frac{0.02 \text{ mol } Na_4Y}{1 \text{ L soln}} \times \frac{416.21 \text{ g } Na_4Y}{1 \text{ mol } Na_4Y}$

$= 8.324 \text{ g } Na_4Y$ target wt.

17

Irv Gratch
10/29/29

Experiment 5: Preparation of 0.02 M EDTA + Zn^{2+} Solns

Purpose: To prepare solns for standardization + metal-EDTA titrations

Procedure - EDTA

1) Weigh ~8.32 g Na_4EDTA into 100-mL beaker on toploading balance, transfer to 600-mL beaker (calc'ns at left)
2) Add ~500 mL deionized (DI) water, stir to dissolve
3) Transfer Na_4Y soln to 1-L plastic bottle
   fill to shoulder w/DI water, cap
4) Invert bottle 20x to mix. Label bottle.

Procedure ~ Zn

5) Weigh 1-cm length Zn wire on analytical balance (use watch glass)
6) Add Zn wire to 600-mL beaker in hood, add 10 mL DI water, add 10 mL HNO_3, swirl till wire dissolves.
7) Rinse 500-mL volumetric flask 3x w/10 mL DI water
8) Transfer Zn^{2+} soln to vol. flask, rinse 3x w/DI water into flask
9) Fill vol flask to shoulder w/DI water, cap, invert 3x, fill to mark w/dropper, cap, invert 20x
10) Rinse 500-mL SC bottle w/~5 mL Zn^{2+} soln 3x (discard rinses), transfer remaining Zn^{2+} soln to bottle, cap
11) Label bottle w/molar concentration Zn^{2+}

EDTA Standardization

You're going to establish the exact concentration of your EDTA solution by taking a known volume of that solution and matching the number of moles of EDTA (symbolized H_2Y^{2-}) in it against the number of moles of zinc in a known sample of the zinc solution. These react on a one-to-one basis:

$$\mathbf{Zn(H_2O)_6^{2+} + H_2Y^{2-} \rightarrow ZnY^{2-} + 2\,H_3O^+ + 4\,H_2O}$$

You'll use a pipet to add the known volume of zinc solution to a flask, then add EDTA solution from your buret until you've added just the right amount to match moles as in the above equation. There has to be a way to know when the right amount has been added; the simplest way to do this is to use a small amount of a compound that changes color when the solution changes from having excess zinc ion to having excess EDTA. Such compounds are called ***indicators***, and since these indicators usually complex the metal ion themselves, different indicators are usually required for different metals. The most commonly used indicator for zinc is a dye called Eriochrome Black T, which is blue in water by itself, but red when complexed to zinc. So a zinc solution with Eriochrome Black T indicator in it will be red and will stay red as EDTA is added—until all the EDTA has taken all the zinc away from the Erio T by forming a more stable complex. Then the Erio T will fairly suddenly turn blue, and the color change signifies that enough EDTA has been added to react mole-for-mole with all the zinc.

Another feature of the reaction equation between hydrated zinc ion and EDTA above is that it generates H_3O^+ along with the zinc-EDTA complex. In water solution, H_3O^+ is the characteristic acid, so the solution is becoming more acidic as the titration proceeds. This is bad, because most of the indicators that are used also change colors with acidity. Since we need to avoid that complication, we control the acidity of the solution by using an additive known as a ***buffer***. A relatively small amount of a buffer solution will control the acidity of the titration mixture, keeping it nearly constant. So for any EDTA titration, it's necessary to know two things—the indicator to be used, and the buffer to be used.

You'll need at least three titrations that match well enough to give you good precision on the determination of the EDTA molarity. But it's hard to be sure you're doing the first one right when you haven't done one before. So we'll want you to begin this time—and every time you do a new *kind* of titration—by starting with a familiarization titration. The instructions below will tell you to use a 250-mL Erlenmeyer flask, and that's what you *should* use for titrations; but for starters, take one of your beakers (250-, 400-, or even 600-mL) and do a titration on a quick-and-dirty basis to see what it's going to look like and roughly where the mole-match will occur. So you're going to take *four* samples: one in a beaker and the other three in flasks. When you do the calculations after the titrations are done, only do them on the three titrations that were done in flasks after you knew what to look for—don't bother calculating up the familiarization titration.

This is another new experiment: Start a new right-hand page and label it "EDTA Standardization". You need the standard subheadings:

- **Purpose (step 14, page 30)**
- **Procedure (step 15)**
- **Apparatus (step 16)**
- **Data (step 17 – tables of titration data)**
- **Calculated Results (step 19 – molar concentration of EDTA, mean, std dev)**
- **Summary and Conclusions**

Use an analytical pipet to transfer 25.00 mL of the standard zinc solution to a clean 250-mL Erlenmeyer flask. Add about 5 mL (use a graduated cylinder) of the NH_3/NH_4^+ buffer solution from the plastic bottle on the shelf. Cautiously sniff the mixture to be sure you can smell ammonia; if not, add more buffer. Take the solution to the instructor, who will add Eriochrome Black T indicator to it. The right amount of indicator will produce a pastel color, not a strong or deep color. Fill the **clean** buret with your EDTA solution after rinsing it three times with small portions of the EDTA solution. Read the initial volume and record it in your notebook in a tabular format (which you should have set up before coming to lab). Begin adding EDTA to the zinc solution. Since both solutions are around 0.02 M, the equal-moles condition means that about 25 mL of the EDTA solution will be required to titrate the zinc—so you can add the EDTA fairly fast at first, swirling the flask (without splashing!) to keep it well mixed. Watch for the formation of a little blue puddle in the middle of the pink solution where the EDTA is going in. Near the equal-moles point in the titration reaction—called the ***end-point***—one drop of EDTA will produce a blue spot as big as a half-dollar. At the end-point, of course, the whole solution turns a clear sky blue when you swirl it, with no purple cast remaining. If you're in doubt about the color, look at it against a window and daylight instead of the lab fluorescent lights. When the end-point has been reached, record in your notebook the volume of EDTA indicated on the buret, to the nearest 0.02 mL; if it's exactly 24 mL, write 24.00 mL, not 24 or 24.0.

The difference between the two buret readings in your notebook is the volume of EDTA solution delivered by the buret. Calculate the molarity (moles EDTA per liter solution) by doing the following dimensional analysis problem (and think about the mole conversion that's going on):

$$?\ \frac{mol\,EDTA}{1\,L\,EDTA} = \frac{0.02500\,L\,Zn}{1\,titration} \times \frac{0.0MMMM\,mol\,Zn}{1\,L\,Zn} \times \frac{1\,mol\,EDTA}{1\,mol\,Zn} \times \frac{1\,titration}{0.0VVVV\,L\,EDTA}$$

In this equation, *0.0MMMM* is the molar concentration of your standard zinc solution to four sig figs and *0.0VVVV* is the volume of EDTA delivered by the buret at the endpoint, expressed in liters (again to four sig figs, meaning five decimals). Keep the leftover zinc solution for now.

Experimental Statistics – ***are your molarity values good enough?***

Okay, now you have three experimental values of M, the molar concentration of the EDTA solution. They all have four sig figs, meaning five decimals, and they should be pretty close together, around 0.02000 somewhere. How close is close? Are your values good enough to rely on, or do you maybe need to run another trial or two? And, for that matter, even if they are really close, which value do you use in later calculations?

Principle 1: *The mean of your experimental values is the best version of the unknown true value.* So calculate a mean value using the printing calculator in the lab in its statistics mode, and note that in your "Calculated Results".

Principle 2: *If you only have random error, the standard deviation of your experimental values is the best guide to the size of your error – but only if you express it as a fraction of your mean.* So calculate the standard deviation of your M values on the printing calculator and note that result in your "Calculated Results"; **then** get what we can call the *relative standard deviation* by getting the standard deviation as a parts-per-thousand fraction of your mean:

$$relative\ standard\ deviation\ (RSD) = \frac{standard\ deviation\ \sigma}{mean\ M} \times 1000$$

Once you have this RSD value, what does it mean? Your individual measurements should have been made and recorded to about 1ppt; that is, a weight might be 0.1884 g, which on the analytical balance is good to about 2 parts in 1884 or about 1ppt, and your endpoint value might be 32.46 mL, which on your buret is good to about 2 parts in 3246 or about 1 part in 1600 or 0.6 ppt. So your overall value *could* be good to 1 ppt, which for a molarity around 0.02 M would give a standard deviation of 0.00002 M and a RSD of 1 ppt. But in fact, experimental errors propagate through your work and your calculations so that, usually, you won't get 1 ppt RSD. Practically speaking, you should be able to get an RSD of 3 ppt – for your *M* value, that would be a standard deviation of 0.00006 M. If your RSD is bigger than 3 ppt, look at your data to see if you can use the Q test to throw a weird value out. In any case, you'll probably need to run another trial or two. After getting new values or discarding old ones, calculate a new mean, standard deviation, and RSD.

Metal-EDTA Titrations

Once you've done at least three titrations (not counting the familiarization titration) of your EDTA solution against your zinc solution that match to good precision (around 3 ppt or less), you're ready to use the standardized EDTA solution to determine the metal content of your coordination compound by another series of titrations. This is a new experiment; start a new notebook page titled "Metal-EDTA Titrations". Use the headings

- **Purpose (step 14, page 30)**
- **Procedure (step 15)**
- **Apparatus (step 16)**
- **Data (step 17 — tables of titration data)**
- **Calculated Results (step 19 — wt % metal in dry compound, mean, std dev)**
- **Summary and Conclusions (step 20)**

But you need to master the quantitative technique on something cheap first, rather than using up all your carefully prepared coordination compound learning how to do the titration. Run a known! Check out a

Table I: EDTA Titration Conditions

Metal	Water/CC Sample	Indicator	Buffer	Solution Conditions
Mn^{2+}	100 mL per 0.2 g	Eriochrome T (***r*** to ***bl***)	5 mL prepared NH_4^+/NH_3 soln	add pinch ascorbic acid before buffer
Fe^{2+}	200 mL per 0.2 g	Xylenol Orange (***r*** to ***y***)	10 drops HCl, 0.5 g hexamethylene-tetramine	add pinch ascorbic acid before buffer
Fe^{3+}	200 mL per 0.2 g	Tiron (***bl*** to ***y***)	10 drops HNO_3, 0.5 g glycine	warm to 40° C
Co^{2+}	200 mL per 0.2 g	Murexide (***y*** to ***vi***)	10 drops HCl, 8-12 drops NH_4OH to true yellow	add murexide to acid to orange-peach, then NH_4OH dropwise to clear yellow
Ni^{2+}	300 mL per 0.2g	Murexide (***y*** to ***vi***)	10 drops HCl, 8-12 drops NH_4OH to lemon color	add murexide to acid to gold color, then NH_4OH dropwise to clear lemon
Cu^{2+}	200 mL per 0.2 g	PAN (***vi*** to ***grn***)	10 drops HNO_3, 0.5 g NaOAc	add 50 mL ethanol per 100 mL water
Zn^{2+}, Cd^{2+}	100 mL per 0.2 g	Eriochrome T (***r*** to ***bl***)	5 mL prepared NH_4^+/NH_3 soln	no special conditions

small sample of a compound with the correct metal in it from the stockroom—it may be a chloride, nitrate, or sulfate, but it will be something that works for your metal. Find out what the formula is supposed to be, and the molecular weight. ***Set the practice-compound titrations up as a separate experiment, then the titrations on your coordination compound as another experiment, each having the headings above.***

All EDTA titrations have several features in common, and Table I at the left is intended to guide you to workable techniques. Every EDTA titration needs an indicator, a buffer, and may have concentration limits and special solution conditions as well. Usually there are several possible indicators, etc., for any given metal ion, but the methods outlined in Table I work pretty well in our labs. When you've decided on a titration method, assume you're going to start by weighing out 0.2000 g of the practice compound. Calculate what that represents in moles of the practice compound and how much of your standardized EDTA solution should be required to titrate it. If that volume is anywhere in the 25-40 mL range, 0.2 g is an appropriate sample weight; if the volume is outside that range, check with the lab instructor about scaling it up or down. In general, you'll weigh out three samples near 0.2 g (0.18-0.22 g to four decimals) and dissolve each sample in 100 mL distilled water in an Erlenmeyer flask, then add buffer and indicator—but note that Table I has a number of special concentration limits, so that you might have to use 200 to 300 mL for 0.2 g of sample (and that would require a 500-mL flask from the stockroom). Remember too that you should run a familiarization titration first in a beaker—a trial that you're probably not going to calculate up, but will only use to give you a feel for what to expect.

GRAVIMETRIC HALIDE ANALYSIS

It's easiest to start by analyzing for weight per cent halide (chloride or bromide). We will do a ***gravimetric*** analysis, in which we turn all of the halide X into solid insoluble silver halide AgX by adding a small excess of silver ion to a solution of the halide:

$$\mathbf{Ag^{+}_{(aq)} + X^{-}_{(aq)} \rightarrow AgX_{(s)}}$$

If we filter off the AgX, dry it, and weigh it, we can calculate the weight of X atoms from the weight of AgX formed, and can get the weight percent X from that value and the weight of the weighed sample that was dissolved to produce the halide ions.

Again this is a new experiment; start a new page, titled "Gravimetric Halide Determinations", with headings

- **Purpose (step 14, page 30)**
- **Procedure (step 15)**
- **Apparatus (step 16)**
- **Data (step 17 — tables of weight data)**
- **Calculated Results (step 19 — wt % chloride or bromide in dry compound)**
- **Summary and Conclusions (step 20)**

The next two pages in the manual show typical notebook pages for the gravimetric experiments.

Because you only have a limited supply of your coordination compound, you'll again need to try a practice run on the appropriate sodium halide NaX to be sure you have mastered the procedure before you start using up your stuff. Make the NaX analysis a separate experiment from the coordination-compound X analysis. Step 1 in the NaX practice experiment will be the gravimetric analysis of weightpercent Cl in NaCl or weight percent Br in NaBr. These are cheap in good purity, and are available in the lab. Begin by weighing out three samples of NaCl or NaBr (whichever halide is in your coordination compound) that are about 0.2000 g, within about 10% or so. That is, three samples

20

Trial 1: $? g\ Cl = 0.2803\ g\ AgCl \times \frac{1\ mol\ AgCl}{143.32\ g\ AgCl} \times \frac{1\ mol\ Cl}{1\ mol\ AgCl} \times \frac{35.453\ g\ Cl}{1\ mol\ Cl}$

$= 0.06934\ g\ Cl$

$\%\ Cl = \frac{0.06934\ g\ Cl}{0.2878\ g\ cpd} \times 100$

$= 24.09\ \%\ Cl$

```
                 MAIN
        0.28030     ÷
      143.32000     x
       35.45300     =
        0.06934   ***
```

```
                  CLR
        0.06934     ÷
        0.28780     x
      100.00000     =
       24.09312   ***
```

Trial 2: wt Cl = 0.06632 g Cl

% Cl = 24.11 % Cl

```
                  CLR
        0.26810     ÷
      143.32000     x
       35.45300     =
        0.06632   ***
```

```
                  CLR
        0.06632     ÷
        0.27510     x
      100.00000     =
       24.10760   ***
```

Trial 3: wt Cl = 0.07231 g

% Cl = 24.03 % Cl

```
                  CLR
        0.29230     ÷
      143.32000     x
       35.45300     =
        0.07231   ***
```

```
                  CLR
        0.07231     ÷
        0.30090     x
      100.00000     =
       24.03124   ***
```

```
                     CLR
              CLEAR DATA
                    STAT
      24.09000 INPUT
      24.09000       ***
      24.11000 INPUT
      48.20000       ***
      24.03000 INPUT
      72.23000       ***
                    STAT
                    CALC
                    MEAN
      24.07667       ***
                   STDEV
       0.04163       ***
                       ÷
       3.00000         √
       1.73205       ***
                       =
       0.02404       ***
                       ÷
                    MEAN
      24.07667       ***
                       x
   1,000.00000         =
       0.99835       ***
```

21

Irv Gratch
10/8/71

EXPT 6 cont.

Procedure

9) Dry crucibles 60 min in oven ($T \geq 120°C$), cool 10 min in desiccator; weigh.

10) Calculate wt % Cl in solid.

Apparatus: 3 sintered glass crucibles, 3 600-mL beakers, 3 stir rods w/policemen, filter flask, vacuum adapter, wash bottle, desiccator, hot plate, oven, analytical balance

Data:

	Trial 1	Trial 2	Trial 3
Coord cpd sample	0.2818 g	0.2751 g	0.3009 g
Crucible + AgCl	28.9455 g	25.4800 g	28.4934 g
MT crucible	28.6652 g	25.2119 g	28.2011 g
AgCl wt	0.2803 g	0.2681 g	0.2923 g

Calculated Results:

	Trial 1	Trial 2	Trial 3
Expt wt % Cl	24.09 %	24.11 %	24.03 %

Mean % Cl: 24.08 % (theor. 24.08 %)

Std dev: 0.042 %

Std dev of mean: 0.024 %

Relative std dev of mean: 1.00 ppt

weighing 0.2069 g, 0.1884 g, and 0.2177 g are OK. Don't waste your time trying to get exactly 0.2000 g. Use the weighing-by-difference technique explained on the posters over the analytical balances, starting with NaX about 1/4 inch deep in your weighing bottle and weighing each sample into a *clean* 600-mL beaker. Note that, although a variation of 10% in weights is acceptable, you still have to *know* each weight to 1 ppt, which is to say to four decimal places. Make sure the samples are numbered 1, 2, 3 in your notebook and that the 600-mL beakers are numbered 1, 2, 3 on the side so you don't accidentally interchange them!

You're going to weigh the AgX that forms by filtering it through your sintered glass crucibles and weighing the crucibles with the dry AgX in them. This means you need a really reliable empty weight for the dry crucible, so you can get the AgX weight by difference. Clean each crucible by numbering it on the side, then inverting it on the suction flask and vacuum adapter and sucking 2 or 3 mL of ammonium hydroxide through it, then rinsing. When the three crucibles *are* clean and rinsed once with deionized water, place them in your labeled crystallizing dish (put your name on the white label-spot using the Sharpie pen) in the lab oven for at least 90 minutes; make sure the oven thermometer is at least 120°C. Be sure the crucibles are numbered before they go in the oven!

Use your crucible tongs to remove the crucibles from the dish, and place each hot crucible in your desiccator to cool. After cooling for 10 minutes, weigh each *numbered* crucible on the analytical balance, picking up the crucible either with tongs or finger cots. Once they're weighed, you can store them in the desiccator — but don't leave them in there for hours or days *before* weighing them, because they will pick up water vapor weight even in a desiccator.

Dissolve each sample in about 150 mL of distilled water; note that you don't have to measure this volume with real accuracy because you aren't analyzing water—it's just there to keep the solution runny. Add about 1 mL of concentrated nitric acid to each solution, using your small graduated cylinder. Calculate the volume (in mL) of 0.5 M $AgNO_3$ solution that would be needed to react with ***(precipitate)*** all the halide ion in your heaviest sample—it should be between 1 and 10 mL. Add about 10-20% to that volume: if the calculated volume is 7.65 mL, use 8.5 or 9 mL of the Ag^+ solution to be sure of precipitating all the halide ion. Add that volume of Ag^+ solution to each sample solution, and immediately place the sample beakers on a hot plate in the hood in dim light. (AgX is light-sensitive, as in photographic film.) Put a clean stirring rod (but not the rubber policeman) in each beaker, and cover each beaker with a ribbed watch glass.

Heat the beakers with their milky suspensions for 20 or 30 minutes, stirring each one occasionally with its own stirring rod. What you want is for the AgX to coagulate into large soft lumps on the bottom, leaving a nearly clear liquid above it. When this happens, take the beakers off the hot plate and place them in your desk in the dark to cool. This is a good place to leave things overnight if necessary.

Filter the samples (number 1 into crucible 1, etc.) by setting up the suction filtration apparatus just as for your synthesis filtration, but with the sintered glass crucible replacing the Buchner flask. However, you can't use rubber dam here. Pour the solution carefully into the crucible (pouring down the stirring rod helps aim it), and use your squeeze wash bottle with deionized water containing a drop or two of nitric acid to rinse the lumps of AgX into the crucible. This will leave a thin film of AgX as a ring around the inside of the beaker at the original water level; use the rubber policeman to squeegee it down into the bottom of the beaker in loose particles, then rinse them into the crucible. It's critically important to get absolutely every visible trace of AgX out of the beaker and into the crucible!

Wash the loose AgX in each crucible by using a stream of liquid from the wash bottle to agitate the solid, with the suction still on. Do this three times to each AgX sample, then allow it to suck dry. Use your crucible tongs or a Kimwipe to remove the crucible from the suction flask and place it in the crystallizingdish in which you dried the empty crucibles. Place the dish and crucibles back in the oven for at least 90 minutes, then remove them, cool 10 minutes in the desiccator, and weigh as before. The weight difference is the weight of AgX you precipitated; convert this to moles of AgX, then to moles of X atoms, then (using the atomic weight of X) to grams of X atoms. The weight percent X in your sample of NaX (or your sample of coordination compound, later on) is (g X atoms)/(g NaX sample) × 100.

When you're through weighing, clean your crucibles out by gently scraping the solid AgX into the "**Waste AgCl**" bottle in the hood, using your microspatula; then wash them with hot soapy water and your test-tube brush. If stains or bits of AgX remain, put the crucible upside down on your suction flask (bottom up), turn the aspirator suction on, and pour one or two mL of concentrated NH_4OH onto the sintered-glass disk and suck it through; this should dissolve any remaining solid and leave a white sintered-glass disk. Rinse once lightly with deionized water and dry.

The only difference you should encounter in analyzing your coordination compound is that, in calculating how much $AgNO_3$ solution to use, you need to remember that a mole of your coordination compound has 2 or 3 or 4 or some other number of moles of halide present. Watch for that factor in setting up your calculation.

ATOMIC EMISSION SPECTROSCOPY (AES)

The analytical methods you've been using up to this point in the lab are called *wet methods* – they involve chemical reactions of known species in solution, like $Co^{2+} + EDTA^{2-} \rightarrow CoEDTA$. These are still used extensively, but in the last 50 years or so many new *instrumental methods* of analysis have been developed. A particularly powerful instrumental method is *atomic emission spectroscopy*, also called *optical emission spectroscopy* and given the quick abbreviation AES or OES. This is the technique we want to move to now.

This is obviously a new experiment. Start a new page titled "Atomic Emission Determination of (*your metal*)." Again the headings will be

- **Purpose (step 14, page 30)**
- **Procedure (step 15–include quiz below and solution prep)**
- **Apparatus (step 16–include instrument: Agilent 4100 MP-AES spectrometer)**
- **Data (step 17–instrument printout, to be mounted on LHP)**
- **Calculated Results (step 19–wt % metal in dry coordination compound)**
- **Summary and Conclusions (step 20–include brief comparison to wet-method results)**

Now let's look at the basis of atomic emission spectroscopy. What we know about isolated (let's say gas-phase) atoms from quantum mechanics is that any given atom has multiple electrons (okay, not hydrogen) that will arrange themselves around the atom's nucleus in such a way as to have the greatest possible attraction for the positively-charged nucleus, but the least possible repulsion for the other negatively-charged electrons – a balancing act, sort of. There will always be one best possible arrangement of electrons. That arrangement has lower energy than any other arrangement, and the atom will always seek it out; it's called the *ground state* of the atom. Atoms sit around in their ground state, normally.

But there *are* other possible electron arrangements, just with less attraction for the nucleus or more repulsion for the other electrons. These have higher energy and are called *excited states*. Now, one result of quantum mechanics is that all of these states have very precise energy differences from the ground state –

exact energies. And each state energy is dictated by the total attraction to the nucleus, which is different for each element because each element has a different number of protons in its nuclei. So if we could measure the energy to excite an atom's electrons from its ground state to, say, the lowest-energy excited state, we could tell Co from Ni, for example, and even in principle tell how much Co or Ni we had. So there's a possible analytical technique here.

Well, how do we put energy into atoms? A more-or-less obvious technique is to heat them – more on this shortly. Another way is to feed the energy into the atom as *electromagnetic radiation*. This works because any electromagnetic radiation – visible light, infrared, x-rays – carries with it energy proportional to the *frequency* of the radiation (symbol ν, Greek *nu*):

$$E = h \cdot \nu$$

energy = *Planck's constant* × *frequency*

So if we put in the right frequency, we'd be putting in the right energy and the atom would move up in energy from its ground state to its excited state. Then what? The atom will always seek out its ground state, but that will mean getting rid of the extra electronic energy it has when it's in its excited state. The atom does this by simply shining out the energy as radiation, with (of course) the frequency equivalent to that energy. This is the *atomic emission* in the name of this technique. If we can measure the frequency, we can look up in a reference what element that corresponds to, and if we can measure its brightness (or *intensity*) we can tell how much of the element is there.

Let's back up to the practical business of putting energy into atoms. We can do it by getting the atoms into the gas phase (so chemical bonds aren't screwing up the atomic electronic energy states) and shining in the appropriate radiation. A variety of techniques do this, in slightly different ways. But as we suggested earlier, one way would be just to heat the atoms. That's not simple, because for the energies we need the equivalent temperatures are enormous – tens of thousands of Kelvin (remember E=*k*T from the kinetic theory of gases? Well, *k* isn't very big; it's 1.38×10^{-23} J/K). So just an ordinary gas flame won't do it.

What we need is a very high *energy density* in our sample of atoms. If you have a joule of energy in a mole of atoms, that's about 10^{-23} J per atom, which isn't much and corresponds to a modest temperature – but if you have a joule of energy in only a thousand atoms that's about 10^{-3} J per atom, which corresponds to a huge temperature equivalent if T = E/*k*. So the total energy is the same, but the sample with only a few atoms has an enormous energy density. We're going to have to do something like this here, and the way we're going to do it is to use the equivalent of a microwave oven. Remember how you keep the microwave time short if you only have a small amount of food in the oven? We're going to use a very small sample and focus high microwave power on it, which will heat the sample so much that (a) all liquid will vaporize, (b) all chemical bonds in molecules will break leaving bare atoms, and (c) the atoms will ionize so that the gas is all charged particles – a *plasma*. This will be plenty of energy per atom to get every atom in an excited state, and all of them will relax back to their ground states (lose their extra energy) by emitting their characteristic frequencies of radiation, even if radiation absorption *wasn't* how the energy was acquired. Those characteristic frequencies are what we need. The instrument that does this is a *microwave-plasma atomic emission spectrometer*.

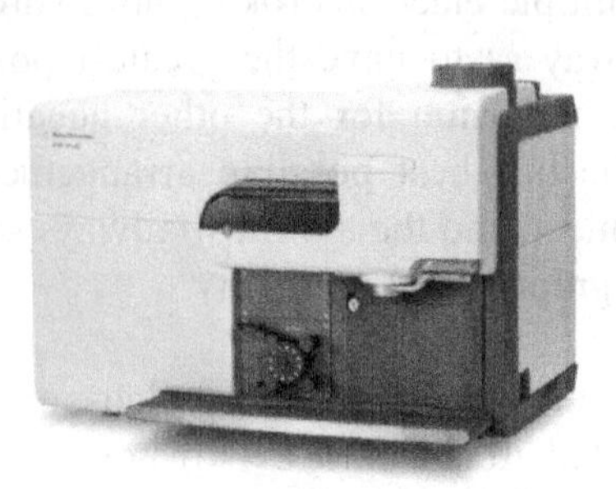

Our instrument is shown at left – an Agilent 4100 MP-AES spectrometer. A liquid sample is sucked into the lower right front of the instrument by a flowing stream of nitrogen gas, and is atomized into tiny droplets as it flows upward into the *plasma*

torch, which is a tube in which microwave energy is focused. The torch raises the energy density in the gas to a temperature equivalent of about 10,000 K, and excites all the atoms. They immediately lose energy by radiating light, which is what we want. The light is focused into a beam, reflected horizontally toward the back of the instrument, then through mirrors that take the light beam to a *diffraction grating*, which spreads the light beam out (like a prism) into rainbow colors. The diffracted beam then travels to the *CCD detector*, which is like the guts of a digital camera – it forms a digital image of the diffracted beam. From that point software takes over. The software knows where on the CCD image to look for, say, Ni, and it can measure the intensity of that particular frequency (see optical diagram below).

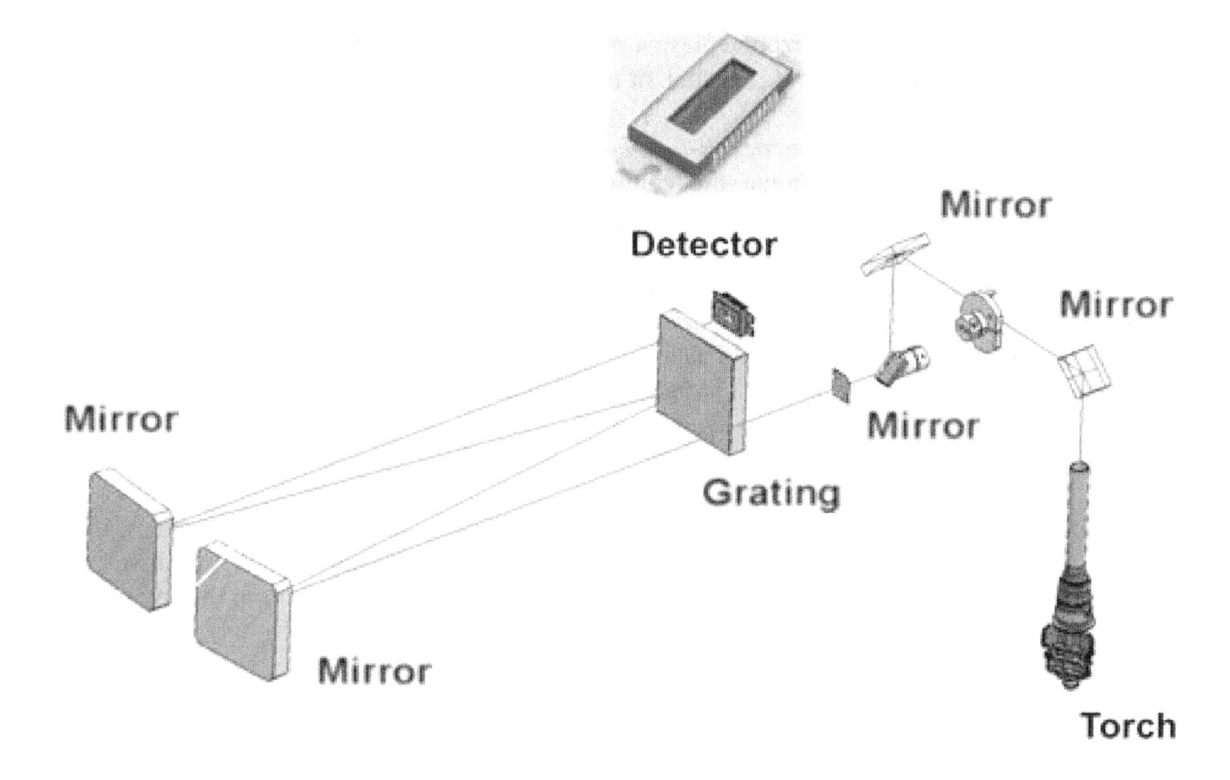

So there are four basic components of the MP-AES:
(1) the ***sample injection*** tube, where a nitrogen flow sucks liquid into the gas stream;
(2) the ***plasma torch***, which atomizes and ionizes the sample;
(3) the ***polychromator***, the system of mirrors and a diffraction grating that turns the light beam into a broad beam of different frequencies (colors); and
(4) the ***CCD detector***, which forms a digital image of the diffracted beam, to be interpreted by the software.

When you've prepared your sample solution (see below) and your standard solution and passed the oral quiz to get your notebook initialed, take both solutions to the instrument room and ask the lab assistant to set up and inject your solutions. In the software, you'll want to look for your metal element.
The results will be reported in *ppm*, parts per million by weight. You can easily convert this number into a weight percentage:

(1) Calculate the ppm of your sample in its solution: You should have weighed out about 0.2 g of your coordination compound, on an analytical balance to four decimals (maybe you weighed out 0.2103 g). You dissolved it and diluted it to 500.0 mL, which weighs 500.0 g. So the ppm for this particular case would be $0.2103/500.0 \times 10^6$, or 420.6 ppm compound in solution.

(2) Now set up the percent calculation for the metal present: just (ppm metal)/(ppm compound) × 100. If your metal is indicated as 88.3 ppm for the solution above, you have 88.3/420.6 × 100 = 20.99 % metal.

Before the operator will set up your solution, he'll need to see initials in your notebook indicating that you've passed the following short oral quiz. The quiz is intended to make sure you have a general handle on how the whole thing works, so it's not just a black (or white) box in the lab.

(1) ***What happens inside an atom when it enters a plasma torch?*** [This involves electron states for the atom, the drying/bond-breaking/ionization/plasma process, the energy-frequency relationship, etc. Must be answered without help from your notes, manual, or notebook.]

(2) ***How does the MP-AES instrument tell us what elements are present and how much light is being emitted by each element?*** [Outline the components of the instrument and the standard/unknown comparison, also without references.]

(3) ***What weight concentrations of your standard metal and of your coordination compound have you prepared, and what ppm metal should the instrument show?*** [This is from your notebook; you should have your solution data, and should also have calculated for the metal what ppm concentration the instrument should show, working from your wet-method percentage.]

Solution Preparation

The AES instrument can always tell that, say, Co is present (or not), but to get an accurate measurement of the *amount* of Co present it needs to be able to compare the light intensity for the "unknown" Co in your coordination compound to the light intensity for a known-concentration Co solution (or any other element, of course). So you'll need to make up a standard solution for your metal.

Metal Standard Solution

You'll need to have a known solution containing about 100.0 ppm of your metal, though anywhere between 2 ppm and 2000 ppm would work. Assume you're going to use your 500-mL volumetric flask, which holds 500.0 grams of water. Now, 100.0 ppm is 100.0/1,000,000 or 1/10,000. So if you weigh out about 0.1 g of pure metal that would be 0.1/500.0 fraction of the volumetric flask contents, by weight, or 2/10,000. That would be 200.0 ppm, which is close enough.

Submit a clean, dry, labeled snapcap vial to the stockroom requesting 0.1 g of your metal for AES. You'll get back pure metal powder or granules. ***On an analytical balance***, weigh by difference a sample near 0.1000 g (± 10%), transferring it into a 250-mL clean beaker. Put the beaker in a fume hood, add 10 mL of concentrated nitric acid (HNO_3), and wait till it completely dissolves–you may need to heat it on a hot plate to get it dissolved. Then transfer the solution to a freshly rinsed 500-mL volumetric flask and rinse the beaker into the flask with three small portions of DI water. Fill the flask to the mark and invert 20×. Calculate the ppm metal in your standard solution (wt metal/500.0 × 1,000,000) and record the concentration in your notebook. Fill one of the two small white plastic bottles in your desk with the standard solution and label the bottle with the metal symbol and its concentration–maybe "Co, 206.0 ppm".

Coordination Compound Solution

Rinse out the 500-mL volumetric flask with DI water 3×. Weigh a sample of your coordination compound near 0.2000 g (± 10%) ***on an analytical balance***, weighing into a clean 250-mL beaker. Add about 25 mL DI water and 1 or 2 mL concentrated HNO_3. Stir (heat if necessary) till it dissolves. Transfer the solution to your 500-mL volumetric flask, rinsing the beaker into the flask 3×. Dilute to the mark and invert 20×. Fill the other small white plastic bottle with the coordination-compound solution and label it with the formula of the coordination compound–maybe "$Co(DMSO)_6Cl_3$".

Instrument Use

Take both bottles of solution to the instrument lab and ask the lab assistant to run your analysis while you watch. He will run your standard *and* a standard the department keeps, for comparison, then run your coordination compound. In the software, select your metal for report.
Print out the report, and mount it on a LHP in your notebook. Calculate the AES values of wt % metal as described above.

FINAL SUMMARY

Your final summary for the entire project should contain a discussion of the synthesis, including your general observations, yield calculation, sources of contamination or error, and possibilities for improvement if any occur to you. There should also be a summary in table form of your analytical results: EDTA standardization, theoretical and observed weight % metal by EDTA, weight % halide, and weight % metal by AES. All observed values except the AA result should have a standard deviation *for the mean* both as an absolute value and as parts per thousand of the mean. On the next page is a version of the table as it might appear in your notebook; note that the table includes the page reference for the set of determinations you're using. If you ran an analysis more than once, give the page reference for the set you want counted.

Use your observed percent halide and the two observed percent metal values to calculate two values of the ***halide:metal mole ratio*** in your compound. Note that this is *not* a weight ratio; convert % values to grams per 100 grams compound and use the atomic weight to get mole amounts. There's an easy theoretical value for mole ratio; in a compound like $FeCl_3(DMSO)_3$ the Cl/Fe ratio is obviously 3.000.

Add some comment on problems you may have encountered in any of the experimental procedures. Be sure to write out the full pledge as it appears on page 8. Notebooks are due on Friday of the last week of lab, as indicated in the schedule at the front of the manual. (The next page has Irv's summary.)

ENDING THE SEMESTER (either by checking out or by preregistering for Ch152)
If you're continuing with Chem 152 lab in the spring, you don't need to check out. You can go right on using the same lab drawer when you get back. On the other hand, if you've only planned to take one semester of lab you *will* need to check out. Either way, your notebook will need to have a slip from the stockroom in the back, or we won't grade it! If you're not going on, check out by the procedure on the last page of the manual; when you finish the stockroom manager will give you a slip saying that you've finished checking out. That slip should be taped inside the back cover of your lab notebook before you turn it in.

On the other hand, if you're planning to continue with Chem 152 lab you should simply take back any extra apparatus you may have checked out from the stockroom and ask the manager for a slip on which you pledge that you're planning to continue with lab. That slip should be taped inside the back cover of your notebook before you turn it in.

87

Irv Grotch
12-7-41

FINAL SUMMARY —

Synthesis (from Jenkins, D.B., J. Am. Chem. Soc. 1971, 93, 4219) proceeded smoothly except that overnight chilling at -20°C in the freezer was required to achieve crystallization. Product was deep purple crystals (7.42 g, 56% yield).

Analytical data —

Det'n	Page Ref Best Data	Theor. Value	Mean Value	Std Dev (single val.)	Std Dev of Mean	Std Dev Mean as ppt
EDTA molarity	27	—	0.01828 M	± 0.000031	± 0.000018	1.4 ppt
% Fe EDTA practice	31	24.64 %	24.39 %	± 0.047 %	± 0.029 %	2.1 ppt
% Fe EDTA coord cpd	37	19.27 %	19.36 %	± 0.051 %	± 0.038 %	2.2 ppt
% Cl grav. practice	45	60.66 %	60.71 %	± 0.14 %	± 0.077 %	0.9 ppt
% Cl grav. coord cpd	49	41.13 %	41.31 %	± 0.092 %	± 0.058 %	2.5 ppt
% Fe MPAES	55	19.27 %	19.14 %	—	—	—
Cl/Fe mol ratio (EDTA Fe)	86	3.000	2.94	—	—	—
Cl/Fe mol ratio (AES Fe)	86	3.000	3.04	—	—	—

SPRING SEMESTER PROJECT: IDENTIFICATION OF ORGANIC COMPOUNDS

In this project you will get a different perspective on chemical techniques from the fall semester, when we used simple synthesis techniques and quantitative analysis techniques to make and characterize a coordination compound whose identity was known (well, more or less). Another very important area of chemical technique involves the identification of unknown organic-compound molecular structures. Over the years many powerful techniques have been developed for this purpose, and we'll introduce a few of these here by identifying some unknown compounds. You'll be given two unknowns: One of them (described on pages 65-122)will be a mixture of two different liquid organic compounds, while the other (pages 122-129)will be a single reasonably pure solid organic compound, a ***carboxylic acid***. During the semester you'll identify all three compounds, partly by physical properties like boiling point or melting point, partly by chemical properties (titrating the acid), and partly—perhaps mostly—by spectroscopic properties. You'll need to keep work on both unknowns going at the same time, and we'll try to help you keep both on schedule.

ORGANIC LIQUID MIXTURE

BOILING POINT AND DISTILLATION (about 90 minutes required; start on notebook p. 5)

The boiling point of a liquid is the temperature at which the vapor pressure of the liquid is equal to the external atmospheric pressure. It's one of the most common and important properties used to identify liquids; extensive tables of boiling points are available. Before beginning the project, however, you may wish to read the section in your lecture text on "***boiling point***" (see its index).

You should divide the project into individual experiments, as you did last semester. This experiment on the liquid will be "Distillation and Separation of Unknown Liquids"; remember the standard headings:

- **Purpose (step 14, page 30)**
- **Procedure (step 15)**
- **Apparatus (step 16 — sketch on left-hand page)**
- **Data (step 17 — temperature ranges in table)**
- **Summary and Conclusions (step 20)**

You will determine the boiling point range of each of your two unknown compounds by a distillation procedure. Besides giving you a helpful value for an important physical property of the compound, the distillation process does two other good things. It separates the two components from each other, since the one with the lower boiling point will distill over first; if you change the receiver in the middle of the distillation, the first receiver will contain mostly the low-boiling component and the second receiver will contain mostly high-boiling component. Also, from the boiling point *range* (the temperature difference between the first few drops of liquid distilling into a given receiver and the last few drops) you can get a crude idea of the purity of the liquid; if the range is only about 2-4 degrees from bottom to top, the liquid's pretty pure, and if the range is 15-20 degrees it's still a fairly impure mixture. So in setting up your notebook for this experiment, be sure you provide a blank table for lowest and highest boiling points of each fraction you distill.

Naturally, you want to separate the two components as completely as possible, because only if each is reasonably pure can you get reliable values of the other physical and spectroscopic properties that will let you identify it. This desire may require you to redistill a particular fraction if it shows a wide boiling-point range or a second large peak (evidence of a second component present) on its gas chromatogram. Consult with the lab instructor if it looks as though that might be necessary.

The distillation apparatus is shown in Figure 2 below. In nearly all techniques for the preparation or handling of small quantities of chemicals, chemists now use standard-taper (Ⓢ) glassware. The principle is obvious on examination of the distillation apparatus; the standardized joints allow various sections of glassware to be assembled for a particular purpose without contamination and without allowing any of the chemical reagents to escape. The joints are sized according to two numbers printed on most of them. For instance, these joints are all Ⓢ 14/20 meaning that the male or inner joint is 14 millimeters in diameter and 20 millimeters long. At the receiver a threaded adapter is used to allow screw-capping the receiver.

Figure 2

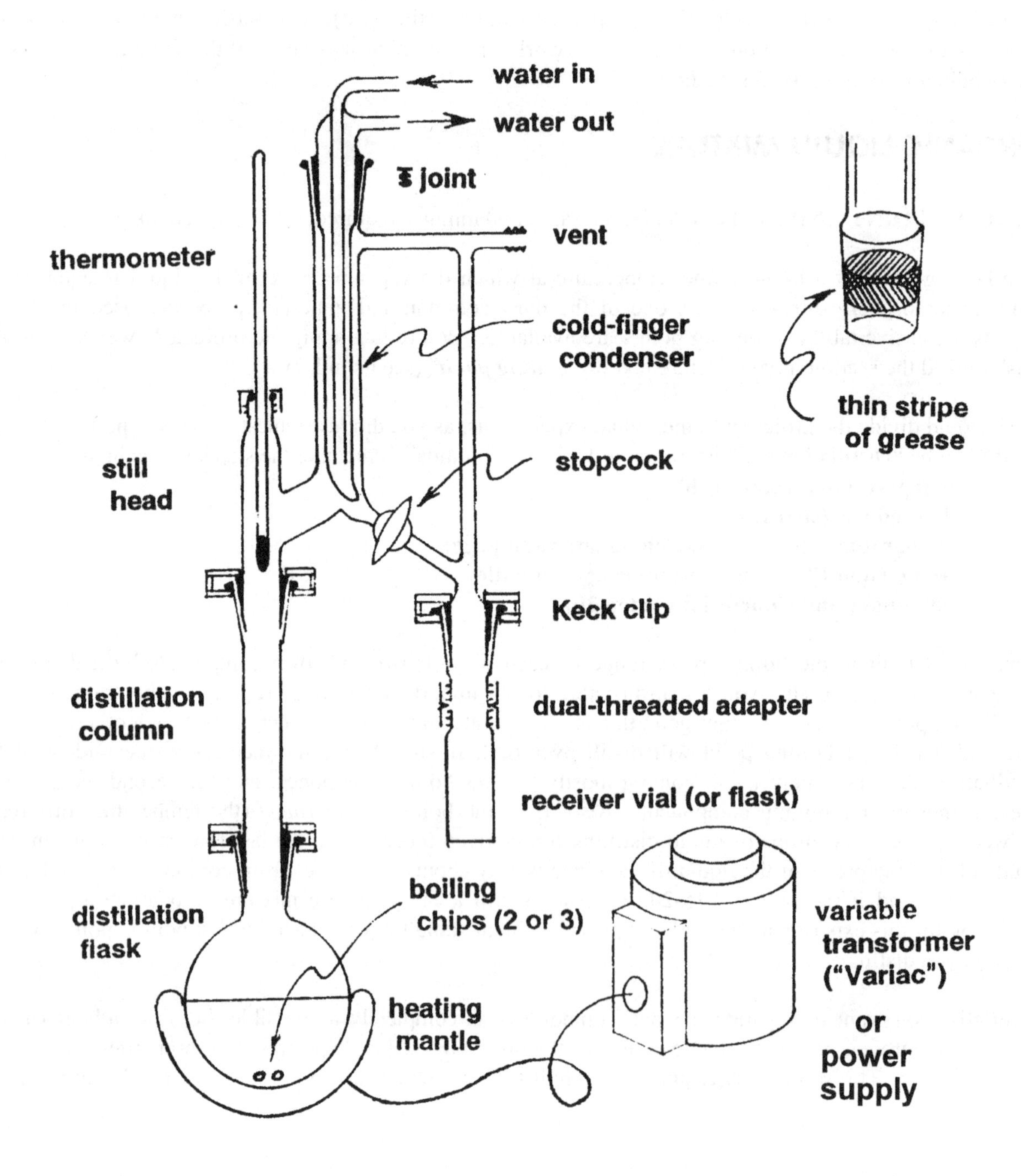

Standard-taper joints must *never* be assembled without a lubricant between the male and female surfaces. Normally stopcock grease is used, which should be applied sparingly as shown in the drawing at left, in a thin stripe around the male joint; properly lubricated areas will be transparent when the joint is assembled. The lubricated area need not cover the full height of the joint, but it must run all the way around.

Study the configuration of the distillation apparatus before beginning work. The ***power supply*** is used to control the power fed to the electric ***heating mantle*** (same idea as an electric blanket). The liquid to be distilled is placed in a ***round-bottom flask*** which the mantle closely fits, and is heated until it begins to boil. The vapor rises up the ***distillation column*** with some of it condensing to a liquid as it goes. The extensive contact between liquid and vapor helps ensure that the *more volatile* component of the liquid mixture distills over first (though the distillate is always a mixture). Thus the vapor coming through the ***still head*** is richer in the more volatile component of the mixture than the mixture itself is. The thermometer, of course, allows us to record the temperature at which the mixture is boiling—the ***boiling point***. The ***cold-finger condenser*** cools the vapor until it condenses into a liquid again, which drips off the bottom of the cold finger. Depending on which way the little drip-tip is pointed, the liquid can all run back into the original flask (often called the ***pot***), in which case you've simply established a liquid-vapor equilibrium at the boiling point, or it can drip off toward the stopcock, so that all the liquid runs out toward the ***receiver*** flask or vial. Note that the cold water always runs into the bottom of the cold-finger condenser and out the top.

The net result is that we get separate ***distillation fractions*** or ***cuts***, with the lowest-boiling-point material coming over first, the highest-boiling last. We can check the purity of these cuts on the gas chromatograph, as discussed in the next section.

Before beginning, make sure that the distillation flask and column are clean and dry, so that no additional impurities soak into your sample. Pour all but a few drops (which should be saved back for a gas chromatogram) of your unknown sample into the lower round-bottom flask or pot; use a funnel (but not your plastic funnel!) to avoid spilling or wetting the joint walls with unknown. The pot should be not more than half full. Add one or two porcelain ***boiling chips***, which provide tiny air bubbles and pores that help nucleate bubbles and boiling. Grease the male joint with stopcock grease as shown in the apparatus sketch. Attach the pot to the distillation column joint with a ***Keck clip***, and raise the clamp holding the heating mantle so that it holds the pot in place.

Check the position of the thermometer in the top of the still head (through the rubber O ring fitting). The *top* of the thermometer bulb should be at the *bottom* of the side arm of the still head. This position is important, because it will affect the boiling temperatures to be recorded. Make sure your receiver vials are clean and dry, and attach one to the still head's dual-screw-threaded adapter by rotating *only the bottom half* of the plastic fitting. Turn the water on to the condenser so that a steady, fairly rapid flow is coming out the outlet tubing into the sink.

After making sure all the joints in the system are tight, turn the switch of the power supply ON, and set the voltage at **5** (about 50 volts) initially. This will have to be increased as the distillation proceeds, depending on the temperatures involved. Consult the instructor before raising the setting above **8**, however.

As the mantle warms the liquid, watch for a line of vapor condensing as liquid on the walls of the pot and rising up the column. When boiling begins, the vapor should rise up the column to the still head fairly rapidly. Initially the thermometer will read room temperature, but as soon as the hot vapor touches it the indicated temperature will rise sharply and more or less level out at the boiling point of the most volatile (lowest-boiling) liquid present, as indicated in the graph on the next page. Don't record the "*Initial T*" for cut 1 until a drop actually falls into the vial, which will be 30 seconds or a minute later.

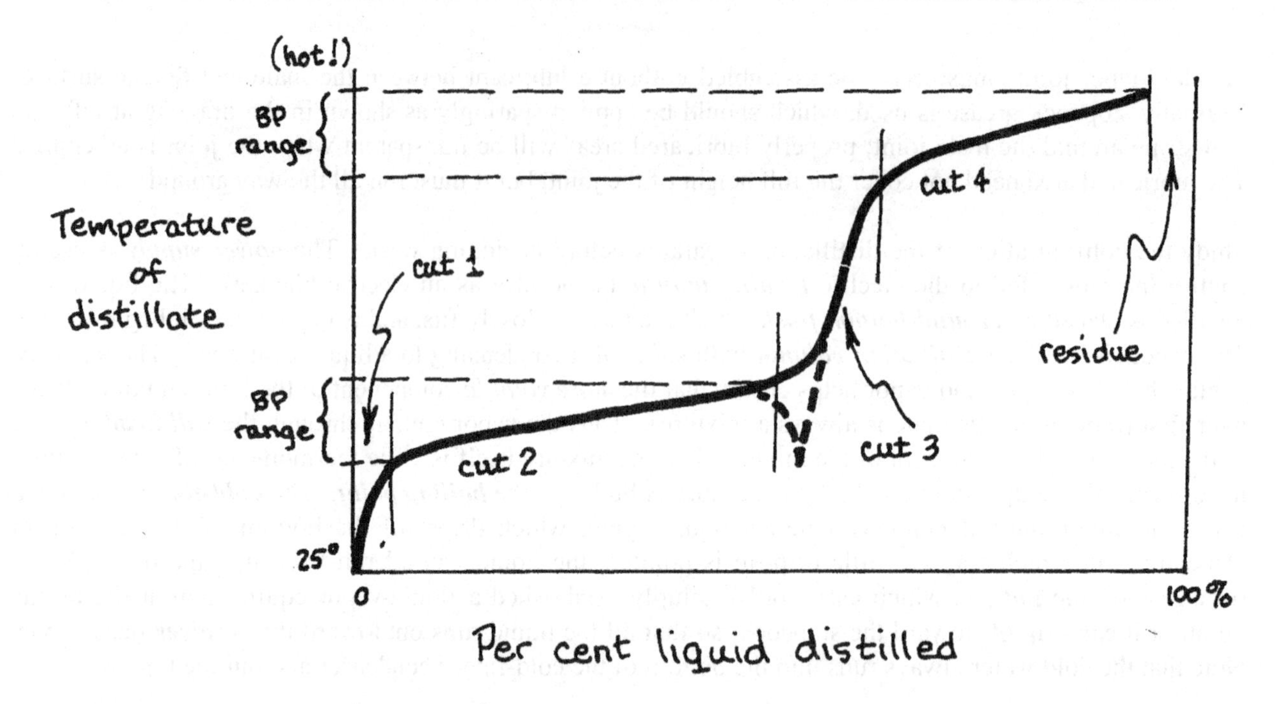

When the temperature seems to have leveled out, which could be anywhere from a few drops in the receiver to a couple of milliliters, you should close the still-head stopcock and remove the receiver vial. Record the temperature in your notebook as "*Final T*" for cut 1. Cap the vial tightly and label it "Cut 1". Attach another receiver flask, open the still-head stopcock, and collect the next distillate as fairly pure low-boiling compound until

(1) the temperature begins to rise again as in the graph above, or
(2) distillation stops and the temperature drops (which means you aren't putting enough heat in to drive over the high-boiling compound), or
(3) you've distilled over about 15 or so mL and are still getting only a slow temperature rise.

The graph above represents the approximate temperature behavior you can expect during the distillation. The heavy line, with its dotted alternative, represents the thermometer reading at the top of the column, and the vertical lines represent the desirable places to change containers for the distilled liquid in the receiver (separate ***cuts***). You should be about at the second vertical line at this point.

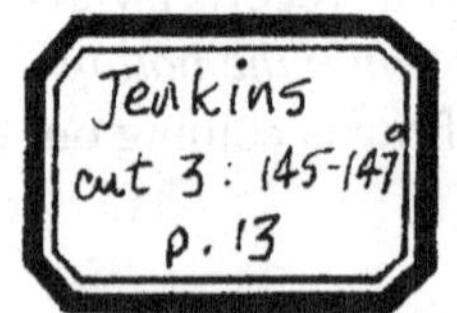

Whichever of the three possibilities you've reached, close the stopcock, remove the receiver vial (which should contain about 15-18 mL, at least an inch deep in the receiver vial), cap it, and label it "Cut 2", including your name, the temperature range, and notebook page reference as in the example at the left. *Cap it tightly!*

Record the temperature range in your notebook as "Cut 2 BP range x - y °C", where x is the temperature at which you switched from cut 1 and y is the highest temperature at which any of cut 2 came over. Attach a new receiver vial to the still head, open the stopcock, and collect the liquid, a mixture, that comes over with a rapidly increasing temperature. Note that if distillation stopped and the temperature went down, you'll have to raise the voltage to the heating mantle to get distillation started again. When the temperature more or less levels out again (should be only a few mL at most), close the stopcock, remove the vial, cap it tightly, and label it "Cut 3" as above. Record the cut 3 BP range in your notebook as you did for cut 2.

Attach another receiver vial, collect the next distillate as fairly pure high-boiling compound until distillation stops or until there are only a few drops of liquid left in the pot—*don't let it boil dry!* Switch off power to the variable transformer so the heating will stop. Remove the receiver vial, cap tightly, and label it "Cut 4" as above. Record the temperature range in your notebook as you did for the previous cuts. When the pot has cooled, pour the three or four mL of undistilled residue into a test-tube, cork it, and label it "Distillation Residue" as above.

Now stop for a minute and think about what your pattern of boiling points looks like. If the small first cut has a very narrow BP range—only a degree or so—and it's the same as most of the second cut, they're probably both pretty pure low-boiling compound. You could combine them, but don't do it until you have a gas chromatogram on each individually (the next experiment!). If the second and/or fourth cuts show only a narrow range of temperature—maybe three or four degrees each—they're probably pretty pure low-boiling and high-boiling components, respectively. If the range of either is much wider than that, you might need to redistill the wide-range component (all by itself) to clean it up. But, again, don't do it until you have a gas chromatogram on each cut to give you an idea of its degree of purity.

GAS CHROMATOGRAPHY AND PURITY (about 20 minutes required; start on notebook p. 15)

As was indicated above, it is highly desirable to have a reasonably quantitative idea of the concentrations of impurities in a given liquid, since these impurities affect the numerical values determined for the various physical properties. If the liquid is quite pure—say 99% or better—the values determined should be in excellent agreement with the tabulated values. But if the liquid contains several percent impurities, one would naturally expect significant deviations from the values tabulated for the pure material. For the purpose of establishing the degree of purity of a liquid sample, the chemist normally uses an instrument called a ***gas chromatograph*** (GC).

Figure 3

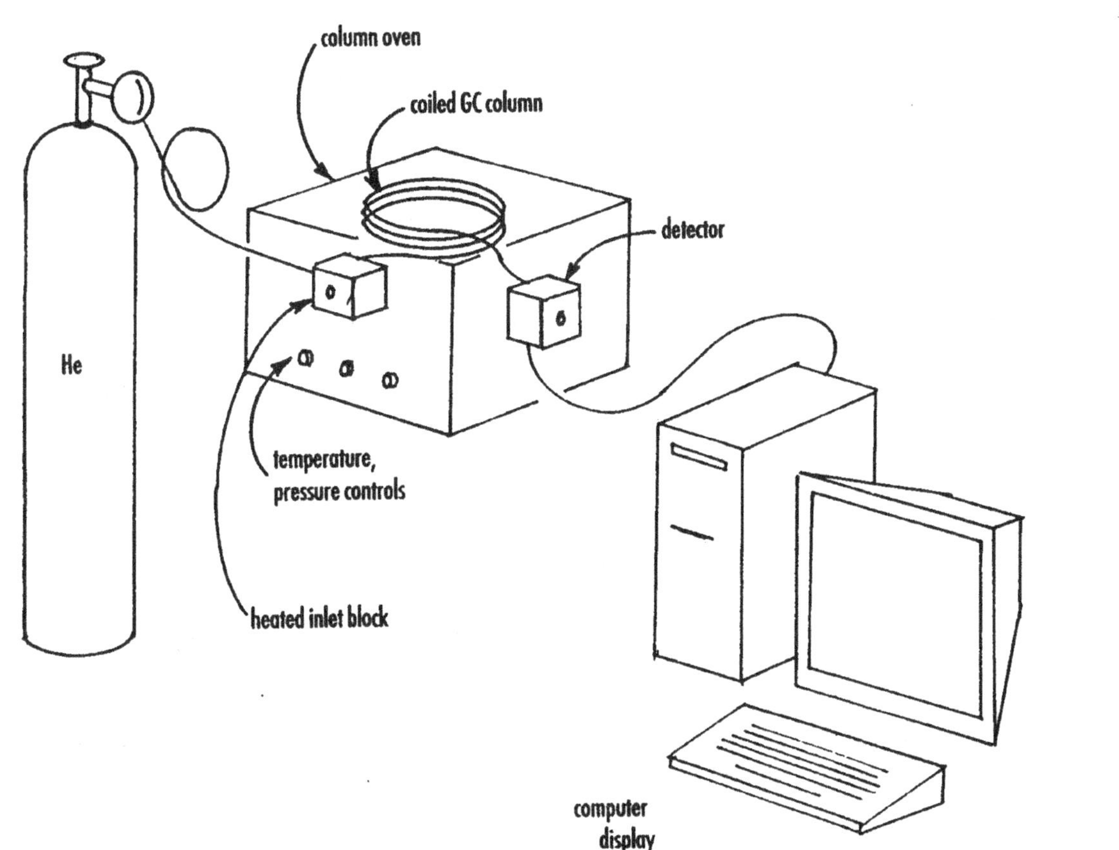

The essential components of a gas chromatograph are shown in Figure 3 on the previous page. The sample is injected through a rubber ***septum*** into the front end of the heated ***column*** where it is vaporized. The sample vapor is then swept down the column by the chemically inert ***carrier gas*** (helium in our case). If the experimental conditions have been properly chosen, some compounds in the sample are held back on the column longer than others so that the various components of the sample are swept out of the column ("***eluted***") one after another. The sample coming off the column is detected by any of several methods. The detector response is recorded as a function of time giving a series of peaks; the printed chart is called a ***gas chromatogram***.

There are basically two kinds of column for gas chromatography. One kind is called a ***packed column***; it's a piece of 1/16 inch stainless steel tubing, several feet long. The column is tightly packed with a finely divided solid (commonly crushed firebrick) which has been coated lightly with a nonvolatile liquid. The other kind of column avoids the solid packing by using very small diameter tubing coated with the nonvolatile liquid; it's called a ***capillary column***. The separating power of the column depends in large part on the nature of the nonvolatile liquid ***stationary phase***.

Upon entering the column, the sample vapor is faced with a choice between remaining in the gas phase that is traveling down the column, or dissolving in the liquid that's sitting still (the stationary phase). Compounds that are very volatile, or that aren't strongly attracted to the particular liquid stationary phase, spend most of their time in the gas phase, and move down the column rapidly. Less volatile compounds spend a larger fraction of the time dissolved in the stationary phase, and move down the column more slowly. Of course, any given molecule spends part of its time moving with the carrier gas, and part standing still with the liquid phase; it's the fraction of the time spent in the gas phase that determines the average velocity down the column. If the column is long enough, and the other experimental conditions are correctly chosen, the various components of a liquid mixture can be separated because of their different velocities through the column.

Note that it is important to select the column temperature carefully. If the column is too cold, the sample will condense and may never come off. If the column is too hot, the sample will be blown straight through the column and no separation will be obtained. Note further that the solubility of a given material in the stationary phase depends on the nature of the stationary phase as well as the temperature. A useful rule of thumb is "like dissolves like." Thus a polar sample would be held longer if the stationary liquid were also polar. This all goes back to the Henry's Law discussion of liquid-vapor equilibrium in your textbook, which you might want to look at. Fortunately, our chromatograph allows us to "program" the temperature to rise during the flow of the sample through the column, which gives us more flexibility.

The detector on the instrument you will use is a ***thermal conductivity detecto***r. In its simplest form this detector is simply a filament, heated electrically and placed in a gas stream. As the composition and hence thermal conductivity of the gas is changed, the temperature and thus resistance of the filament changes. The instrument balances the filament resistance against another adjustable resistor in a ***bridge*** to get zero net current flow, but when the temperature changes the bridge becomes unbalanced, and the resulting current produces a signal that the controlling computer displays as a peak.

The area under the peak produced by a thermal conductivity detector is proportional to the amount of the compound being detected. If all organic compounds had the same thermal conductivity, it would be a simple matter to measure the relative concentrations of the components of a mixture. Thermal conductivity *does* vary from compound to compound, but not greatly. Thus ballpark estimates of concentrations can readily be made. If an impurity is 10% as big as a sample peak, then the impurity is probably between 5% and 20% of the sample, i.e., 10% ballpark.

This is another experiment; you need headings

- **Purpose (step 14, page 30)**
- **Procedure (step 15)**
- **Apparatus (step 16 — include instrument model: SRI 8610B gas chromatograph)**
- **Data (step 17 —instrument printouts taped in on left-hand page)**
- **Summary and Conclusions (step 20)**

Once you have the boiling points of your compounds, consult the gas chromatograph operator to get a time during the laboratory period when you can run your series of liquids. At that time, take the operator a sample of your *original liquid mixture* and samples of *each cut* from your distillation and watch the chromatographing of your precious samples. Mount the chromatograms on left-hand pages in your notebook (permanently, with Scotch tape), and comment on the opposite right-hand pages on the relative purity of your samples before and after distillation.

SIEVING THE TABLE FOR MOLECULAR STRUCTURES (start on notebook p. 20, both sides)

Now you have an approximate value for the boiling point of each of your two unknown liquids, assuming the GC showed that each is pretty pure—99% or better. It's time to go to the table at the end of the "Spectroscopic Properties" discussion and pick out the possibilities for each compound. Obviously, the thing to do is to look through the table and pick out the compounds with matching boiling points, but it isn't quite that simple. There are two potential problems that have to be allowed for. First, you're not *sure* either of your liquids is absolutely pure, and that would of course affect your boiling points. Second, you probably didn't do the distillation at the same barometric pressure as the reference handbooks report, and that will affect the BP at least a little bit. The way to allow for these potential problems is to use a *wider* range of BP's than you actually observed when you go through the table. In general, you should probably note as a low-boiling possibility anything that has a listed BP from about 8 degrees below the bottom of your range to about 2 degrees above the top of your range. That is, if you observed 105-109°C BP range (a four-degree range), you should consider anything from 105 - 8 or 97°C to 109 + 2 or 111°C. On the other hand, for your high-boiling component, consider anything from 2 degrees below the bottom of your range to about 8 degrees above the top of your range: for example, if you saw 144-146°C, consider anything between 142 and 154°C.

So, still under the general distillation heading in your notebook, go through the table and make two lists: one for candidates (on the basis of BP) for your low-boiling unknown compound and one for candidates for your high-boiling unknown. **Include the numeral in column "C" of the table for each compound.** Be thorough, because those lists will be the basis for a number of weeks' work from here on.

SPECTROSCOPIC PROPERTIES—INTRODUCTION

At this point in the project you have established the boiling point range for each compound and established a short list of hot prospects from the table for each of your compounds; that is, you should have found somewhere between about ten and thirty candidates for each, all with about the right boiling point. Basically, that's almost as far as organic chemists usually try to go with bulk physical properties. It's possible to run chemical reactions and make derivative compounds, then look at their physical properties, but that's pretty tedious—and it doesn't get at the real difference between the various candidates, which is their different molecular structures.

The spectroscopic techniques available to chemists today allow us to distinguish compounds from each other on the basis of their specific molecular frameworks. During the rest of the project, you'll be looking at different spectroscopic techniques as they apply to your compounds: you'll find that they give very detailed, very specific information about the molecular structures of your compounds as you integrate together the various pieces of information from each technique for a given compound.

What we need first is to understand in general what spectroscopic techniques do for a molecule. *Spectro-* derives from a Latin word meaning form or appearance, and *-scope* derives from a word meaning to see or aim at. So we will, almost literally, look at the appearance of light that's been modified by the compound. Now light is electromagnetic radiation, and it has a range of wavelengths or frequencies far greater than the narrow range that our eyes are sensitive to; we won't actually look at visible frequencies of light, but we will look at several other fairly narrow ranges of frequencies. Each of these ranges, as it comes from a molecule, gives us certain kinds of information about the structure of the molecule—its nuclei, electrons, and chemical bonds.

Your unknown compounds are organic molecules. That means that each one is made up of electrically neutral molecules consisting of carbon atoms bonded to each other in some kind of carbon-carbon bond skeleton. Then, because carbon almost always forms four bonds to other atoms and there aren't usually four other carbon atoms to bond to, there are almost always hydrogen atoms present to sop up the bonding capacity of the carbon atoms. If a molecule contains only these atoms, it's a ***hydrocarbon*** molecule, and some of you have hydrocarbon molecules in your unknown mixtures. But most organic molecules have noncarbon, nonhydrogen atoms in them—most commonly oxygen, nitrogen, or a halogen, particularly chlorine or bromine. The patterns of chemical bonds from carbon to these ***heteroatoms*** (different atoms) are called ***functional groups***. For instance, there is a class of compounds containing an oxygen atom with a double bond to a single carbon atom, and two other carbon atoms having single bonds to the C=O; the C=O is a ***carbonyl*** functional group, and all carbonyl compounds having two carbon atoms bonded to the C of the C=O are called ***ketones***. The functional groups we'll be dealing with in these unknowns are indicated in Table III on the next page, and we'll look at where these names come from shortly.

All of this means that if spectroscopy is going to help you identify your unknown compounds, it has to give you three different pieces of information: (1) the nature of the carbon-carbon skeleton of the molecule; (2) the kinds of functional groups present in the molecule (if any); (3) the number and locations of the hydrogen atoms present in the molecule. The spectroscopic techniques we're going to introduce you to over the next six weeks or so will answer these questions in the order we've just mentioned.

First, we'll get most of the information about the carbon skeleton by using ***^{13}C*** (pronounced "carbon thirteen") ***nuclear magnetic resonance*** (NMR) spectroscopy. In this experiment, you'll find out how many symmetrically different kinds of carbon atoms there are in your molecule, different in the sense of having different densities of electrons circulating about them. For instance, ***chlorobutane***, $CH_3CH_2CH_2CH_2Cl$, has four different kinds of carbon atoms: one with a chlorine on it, one next to one with a chlorine, one that's two carbons away from the chlorine, and one that's three carbons away. ^{13}C NMR would show you that there were four kinds of carbon atoms present in the molecule. If the alternative were a molecule with only three carbon atoms, you'd know right away to disregard the three-carbon candidate.

Second, we'll find out about the presence or absence of heteroatom functional groups like those in the table on the next page by using ***infrared spectroscopy***. In this experiment, you'll find out whether there are groups like C=O, C-O-C, C-OH, or C-NH_2 present in your molecule (and note that it's just as important to know that those functional groups are *not* there as it is to know they *are* there). Our infrared spectrophotometer also has a large library of infrared spectra, for essentially every molecule you could have, and once you have information on the carbon skeleton and the functional groups that are present you can do a computer search of the infrared spectra data base to find spectra to compare to the one you've just gotten.

Third, we'll find out about the number and location on the molecule of the hydrogen atoms that are present, by using ***^{1}H*** (call it "proton") ***NMR*** spectroscopy. Like the ^{13}C NMR, proton NMR will tell you the different electron densities around each kind of hydrogen atom, but it will also give you the relative number of each kind of hydrogen atom, and will even give you the number of hydrogen atoms on the next-neighbor carbon atom in the carbon skeleton.

The end result of all this is an overall pattern of spectroscopic information about your compounds that you can put together with the boiling point (and possibly other physical properties) to give you a conclusive identification of each. These are very powerful techniques, but they're also pretty sophisticated. We don't

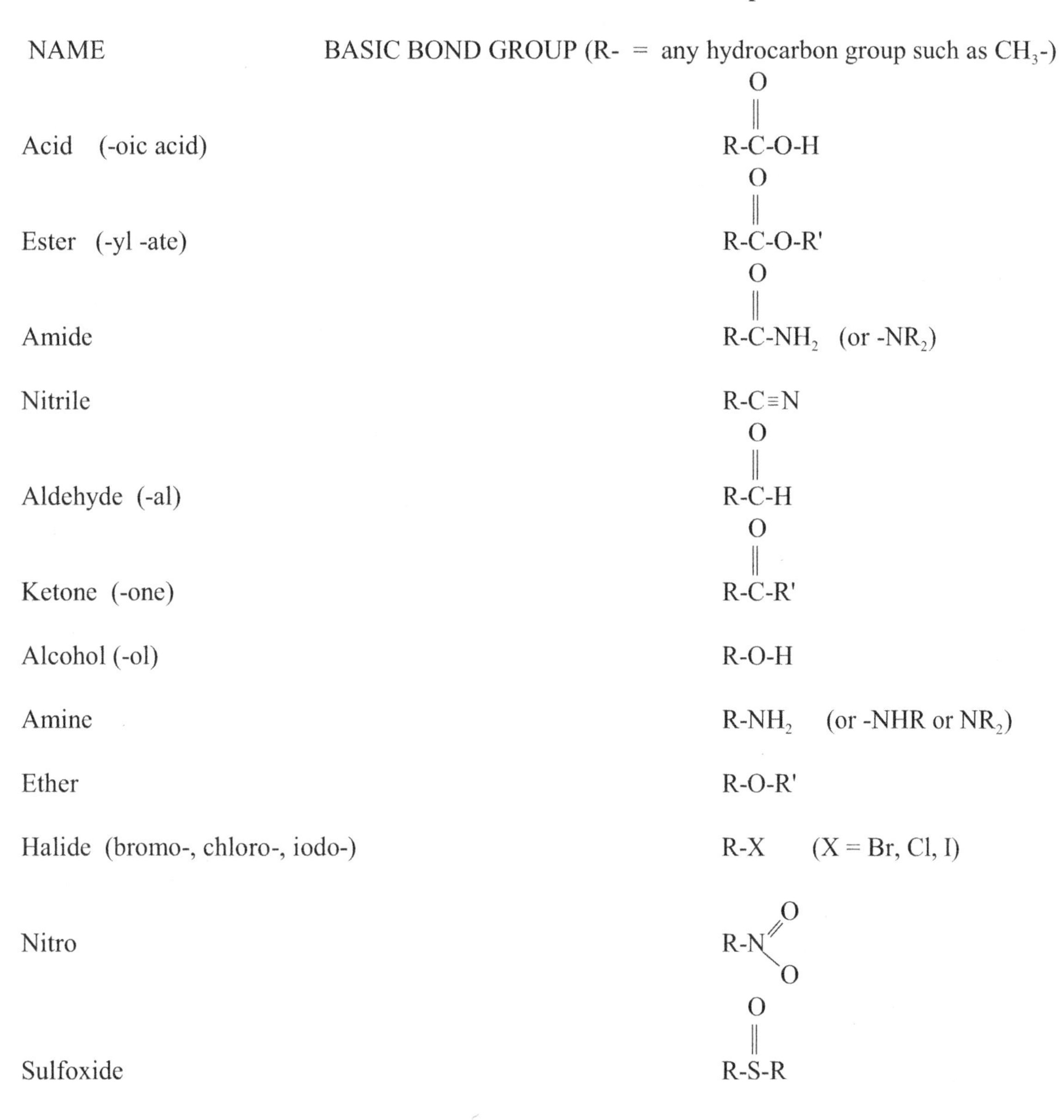

Table III: Functional Groups

NAME	BASIC BOND GROUP (R- = any hydrocarbon group such as CH_3-)
Acid (-oic acid)	R-C(=O)-O-H
Ester (-yl -ate)	R-C(=O)-O-R'
Amide	R-C(=O)-NH_2 (or -NR_2)
Nitrile	R-C≡N
Aldehyde (-al)	R-C(=O)-H
Ketone (-one)	R-C(=O)-R'
Alcohol (-ol)	R-O-H
Amine	R-NH_2 (or -NHR or NR_2)
Ether	R-O-R'
Halide (bromo-, chloro-, iodo-)	R-X (X = Br, Cl, I)
Nitro	R-N(=O)-O
Sulfoxide	R-S(=O)-R

want you to think of each of them as just another mysterious black box. So over the next twenty pages or so (and in the lab lectures) we'll explain how each one of them works. For each technique, when you've studied the manual and your lecture notes (and gotten questions answered by the faculty or the lab assistants!), there will be a short quiz (like the AA quiz) on what the technique tells you and how it works. At the end of each of the sections that follow, we'll tell you what the questions on the quiz are going to be. When you've figured out the answers, come in and tell a faculty member you want to take the quiz; we'll give it to you on the spot, and if you run into trouble with it we'll let you go study some more and take it again with no penalty. Sooner or later you'll pass the quiz, and then we'll initial your lab notebook and let you go run the spectrum yourself.

NAMING ORGANIC COMPOUNDS

You're going to be using spectroscopic techniques to identify your compounds by establishing features of their molecular structures. That means you have to have some feel for what molecular structures can exist for organic compounds. The table of liquid compounds at the end of this section has a lot of these, and before you can use it with its list of names you need to understand the basics of how the names represent the molecular structures. So here's a first pass at ***organic nomenclature***.

Carbon, more than any other element, tends to form compounds by making chains of bonds between similar atoms: -C-C-C-C-, and so on. The essence of organic names is the identification of the longest chain of C atoms, or ring of C atoms, and using a name for that chain as a base for any other atoms that may be there. In the chart on the opposite page, Part I gives the names for unbranched chains of carbon atoms that have satisfied the bonding capability of the carbon atoms by filling out their valence electron groups with hydrogen atoms. A carbon atom normally forms four bonds and hydrogen one, so the number of hydrogen atoms on a given carbon is just four minus the number of other bonds to that carbon. In a chain, the end C is bonded to one other C, so it has three H's attached and it's a CH_3- group (***methyl***). One of the inner C atoms is bonded to two other C's so it has two H's as a -CH_2- group (***methylene***).

But the basic name for the ***hydrocarbon*** is taken from the chain length; note that there are special names for one- through four-carbon chains, but from then on it's just the Greek numeral names. Pentane has five C atoms in its chain just as a pentagon has five sides. If the chain is closed into a ring, we keep the chain name but add the prefix *cyclo-* as in cyclohexane. If there is a C=C double bond in the chain, the *-ane* ending changes to *-ene*, and if there's a C≡C triple bond the ending becomes *-yne*. If we have more than one of a particular bond or heteroatom, the name takes on a Greek prefix: *di-* for two, *tri-* for three, *tetra-* for four. So a four-carbon chain with two double bonds in it would be ***butadiene*** (not dibutene, because that suggests eight carbon atoms).

The name of a hydrocarbon with a double bond in it ought to tell us where the double bond is in the chain. We number the C atoms from one end of the chain so as to get to the double bond *as soon as possible* and give the C atom number where the double bond starts (that'll be the *lowest possible number*). Thus C=C-C-C would be ***1-butene***, but C-C=C-C would be ***2-butene***. Note that C-C-C=C is still 1-butene!

There's a special class of ***aromatic*** hydrocarbons (they have a stronger smell than open chains) that carry their own names because of their special stability: A six-membered ring with three C=C double bonds alternating with three C-C single bonds would be *cyclohexatriene* by the rules we've just developed, but it's called ***benzene***. Part II of the chart on the next page gives a few names for common aromatic hydrocarbons. Note that if two groups are attached to a benzene ring we have to locate them; ***xylene***, which has two methyl groups, can have them next to each other (*1,2-dimethylbenzene* or *ortho-xylene*, abbreviated **o-*xylene***), one carbon apart (*1,3-dimethylbenzene* or *meta-xylene*, **m-*xylene***), or opposite each other (*1,4-dimethylbenzene* or *para-xylene*, **p-*xylene***).

For the ***functional groups*** we've already seen in Table III, we can name their compounds by (1) adding their prefixes or suffixes to the chain name and (2) adding a number to indicate where on the chain the group is located. In Table III the groups are in naming-priority order; that is, if two functional groups are present the higher one in the table should form the basis for the name and the lower one in the table should just be added on. Thus, Cl-CH_2-COOH would be ***chloroacetic acid***, not *carboxylchloromethane*. The two carbons in this compound suggest naming it as *chloroethanoic acid*, which is actually legitimate. However, Part III of the chart gives more commonly used names for the chains in carboxylic acids. These names also are applied in ***esters***, which are named *(alcohol)yl (acid)ate*, such as ***ethyl acetate***, which has a CH_3CH_2- group from *eth*anol and a CH_3COO- group from *acet*ic acid.

Organic Chains of Carbon Atoms and Names

Part I: Names of Unbranched Chains

CH_4	CH_4 = ***methane*** (CH_3- is *methyl*)
CH_3-CH_3	C_2H_6 = ***ethane*** (CH_3CH_2- is *ethyl*)
CH_3-CH_2-CH_3	C_3H_8 = ***propane*** ($CH_3CH_2CH_2$- is *propyl*)
CH_3-CH_2-CH_2-CH_3	C_4H_{10} = ***butane*** ($CH_2CH_2CH_2CH_2$- is *butyl*)
CH_3-CH_2-CH_2-CH_2.CH_3	C_5H_{12} = ***pentane*** (C_5H_{11}- is *pentyl*, or *amyl*)
CH_3-CH_2-CH_2-CH_2-CH_2-CH_3	C_6H_{14} = ***hexane*** (C_6H_{13}- is *hexyl*)
CH_3-CH_2-CH_2-CH_2-CH_2-CH_2-CH_3	C_7H_{16} = ***heptane*** (C_7H_{15}- is *heptyl*)
CH_3-CH_2-CH_2-CH_2-CH_2-CH_2-CH_2-CH_3	C_8H_{18} = ***octane*** (C_8H_{17}- is *octyl*)

Similarly C_9H_{20} is ***nonane***, $C_{10}H_{22}$ is ***decane***, and so on.

Special Names of Some Branched Chains

$(CH_3)_2CH$-	*2-propyl*	but also	***isopropyl***
$(CH_3)_3C$-	*dimethylethyl*	but also	*tertiary butyl* or ***tert-butyl***
$(CH_3)_2CHCH_2$-	*2-methylpropyl*	but also	***isobutyl***
$CH_3CH_2CH(CH_3)$-	*2-butyl*	but also	*secondary butyl* or ***sec-butyl***

Part II: Special Names of Aromatic Ring Systems

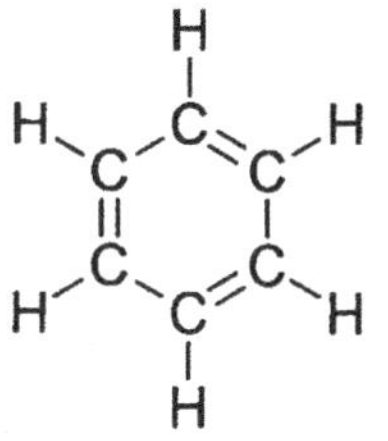

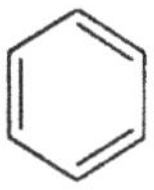

= C_6H_6 = ***benzene*** 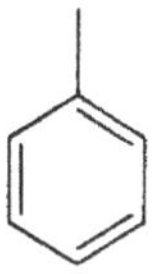 $C_6H_5CH_3$ = ***toluene***

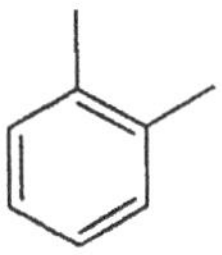

$C_6H_4(CH_3)_2$ = **o-*xylene*** 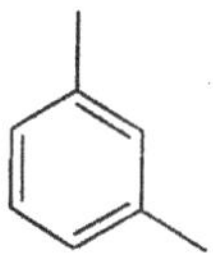 **m-*xylene*** 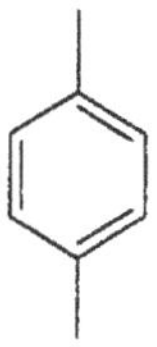 **p-*xylene***

Part III: Special Names of Chains in Carboxylic Acids

H-COOH	C_1	***formic acid***
CH_3-COOH	C_2	***acetic acid***
CH_3-CH_2-COOH	C_3	***propionic acid***
CH_3-CH_2-CH_2-COOH	C_4	***butyric acid***
CH_3-CH_2-CH_2-CH_2-COOH	C_5	***valeric acid***
HOOC-COOH	C_2	***oxalic acid***
HOOC-CH_2-COOH	C_3	***malonic acid***

---

^{13}C NUCLEAR MAGNETIC RESONANCE (NMR) (after quiz, about 30 min.; start on notebook p. 25)

Here's the first spectroscopic technique for determining molecular structure. *Spectroscopy* implies radiation, and it has to interact with molecules somehow. To understand ^{13}C NMR, we'll need to understand two things. **First**, what happens when this particular radiation hits these molecules in the environment they have inside the instrument? **Second**, what are the hardware components of the instrument; how do they generate the radiation, apply it to the molecules, and then analyze the result to tell us something about the molecule?

Magnetic Properties of Molecules

Let's look at the molecules first. The name of the technique, ***magnetic resonance***, suggests that we're going to apply a magnet to the molecules and look at a magnetic property the molecules have. The thing is, most molecules don't have any ***magnetic moment***; they don't interact with a magnetic field. The molecules are made up of atoms bonded together, which means electrons and atomic nuclei sprinkled around through the molecule. Now, electrons have a property called ***spin***, and a spinning electrical charge is just like a electromagnet coil – it has a magnetic moment. But when electron spins are *paired*, they cancel each other out and the *pair* of electrons has no magnetic moment. And in the kinds of molecules we're looking at here, all the electrons are paired. So, nothing to work with as far as electrons are concerned.

Then we're looking at atomic nuclei (which is why it's *nuclear* magnetic resonance). Most atomic nuclei have protons and neutrons paired in much the same way electrons are paired, so although protons and neutrons have individual spins, they don't yield a magnetic moment. And, in fact, the normal abundant isotope of carbon, ^{12}C, has no overall nuclear spin or magnetic moment. But the ^{13}C isotope nucleus, which is about 1% of all natural carbon, *does* have a spin (it has an odd number of nucleons, so they can't all pair their spins) and a ***nuclear magnetic moment***. So these carbon atoms will interact with a magnetic field.

All right, how do these ^{13}C nuclei tell us anything about the molecule's structure? First, even though they're only 1% of all the carbon, there are so many trillions of molecules in even a tiny sample that we can be sure there will be a ^{13}C in every carbon position in some molecule or other. So we can investigate carbon atoms in every atomic position in a molecule.

What are we going to do with these magnetic nuclei in a magnetic field? Well, the nuclear magnetic moments will align themselves parallel to the ***external magnetic field***, just like the needle in a magnetic compass aligns itself with the earth's magnetic field – that is, "points north". So all these magnetic moments are simultaneously aligned parallel to the external magnetic field. What does the radiation have to do with it? Absorbing radiation is how we get *energy* into an atom or molecule, because electromagnetic radiation carries energy with it:

$$\Delta E = h\nu$$

This is Planck's relation: **ΔE** is the energy change that occurs when the radiation is absorbed, ***h*** is Planck's constant, and ν (Greek *nu*) is the *frequency* of the radiation. Radiation carries *energy proportional to its frequency.*

Now, the strength of interaction between a nuclear magnetic moment and the external magnetic field depends on how strong the magnetic field is. The stronger the magnetic field, the stronger the interaction. So the energy of magnetic interaction, $\Delta E_{magnetic}$, is proportional (using a constant γ, Greek *gamma)* to the field strength **H**:

$$\Delta E_{magnetic} = \gamma H$$

The "delta" here means an energy *change*, of course, and the change in this case is the energy difference between the two possible magnetic states. In the low-energy state the nuclear magnetic moment is *parallel* to the magnetic field, and in the high-energy state it's opposed – *antiparallel* – to the magnetic field.

In this case, the energy absorbed when radiation is absorbed is the amount needed to flip the nuclear magnetic moment over so that it's antiparallel to the external field. It's as if we had reached into the magnetic compass and forced the needle to point south, instead of north.

$$\Delta E = h\nu = \Delta E_{magnetic} = \gamma H$$

or just

$$h\nu = \gamma H$$

Now, here comes the tricky part. You might think the individual nuclear magnetic moments would all take the same energy and thus the same frequency to flip over, since they're all in the same external magnetic field. But in fact they're *not* in the same magnetic field! Depending on where each carbon nucleus is in the molecule, there will (or at least might) be a different density of electrons around it. Being near a bromine atom with 35 electrons, for instance, would put the carbon nucleus near a big electron density. And although the paired electrons have no net magnetic moment, they can circulate around within the molecule. The circulation is like electrons flowing through a wire coil – it makes its own little magnetic field, which in this case we refer to as the ***internal magnetic field*** H_{int}. Then the *total* magnetic field each carbon nucleus sees is the sum of the big external magnetic field H_{ext} (the same for all nuclei) and the small internal magnetic field H_{int} (different depending on what the density of circulating electrons is in that part of the molecule.) So we have

$$h\nu = \gamma(H_{ext} + H_{int})$$

Sometimes symmetry can make two or more different ^{13}C nuclei ***magnetically equivalent***, but in general different carbon atoms will require different amounts of energy to flip over the nuclear magnetic moment. That means different frequencies of radiation will be absorbed by different carbon atoms; they're ***magnetically distinct.***

You can imagine that if we could somehow scan all the possible frequencies, and we saw that four different frequencies were absorbed, we could say that the molecule has a structure such that there are four magnetically distinct carbon atoms. That's useful. But we can go farther than that. You could imagine that there would be a particular frequency characteristic of a carbon next to a bromine, for example. Now, we can't tabulate a bunch of absolute frequencies like this because they will be different for different big external magnets. But what we *can* do is to pick a reference compound that has just one kind of carbon, put a little of it into each sample, and express the carbon-next-to-bromine frequency as a ***chemical shift*** from the frequency of the reference compound. The chemical shift can then be a fraction; in our case, not a percentage because it's really small, but a fraction expressed in ***parts per million***.

The reference compound for ^{13}C NMR is ***tetramethylsilane***, or **TMS**:

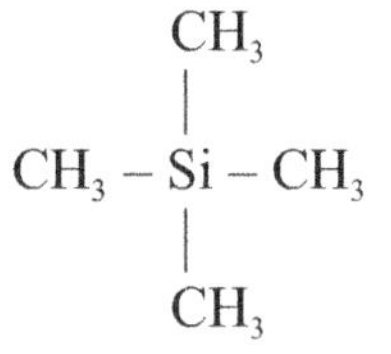

```
        CH3
         |
  CH3 – Si – CH3
         |
        CH3
```

(C atoms are at the corners of a tetrahedron around Si)

You can see from its symmetry that although it has four carbon atoms, they're all in exactly the same electronic environment, so they'll all absorb the same frequency.

For our instrument and our magnet, the frequencies absorbed by ^{13}C nuclear magnetic moments flipping will be somewhere near 100 MHz – about the frequency of FM radio. All of the chemical shifts will be somewhere between 0 and about 220 parts per million, or ***ppm***. Table IV below gives general ranges for the

Table IV—^{13}C NMR Chemical Shifts vs. TMS

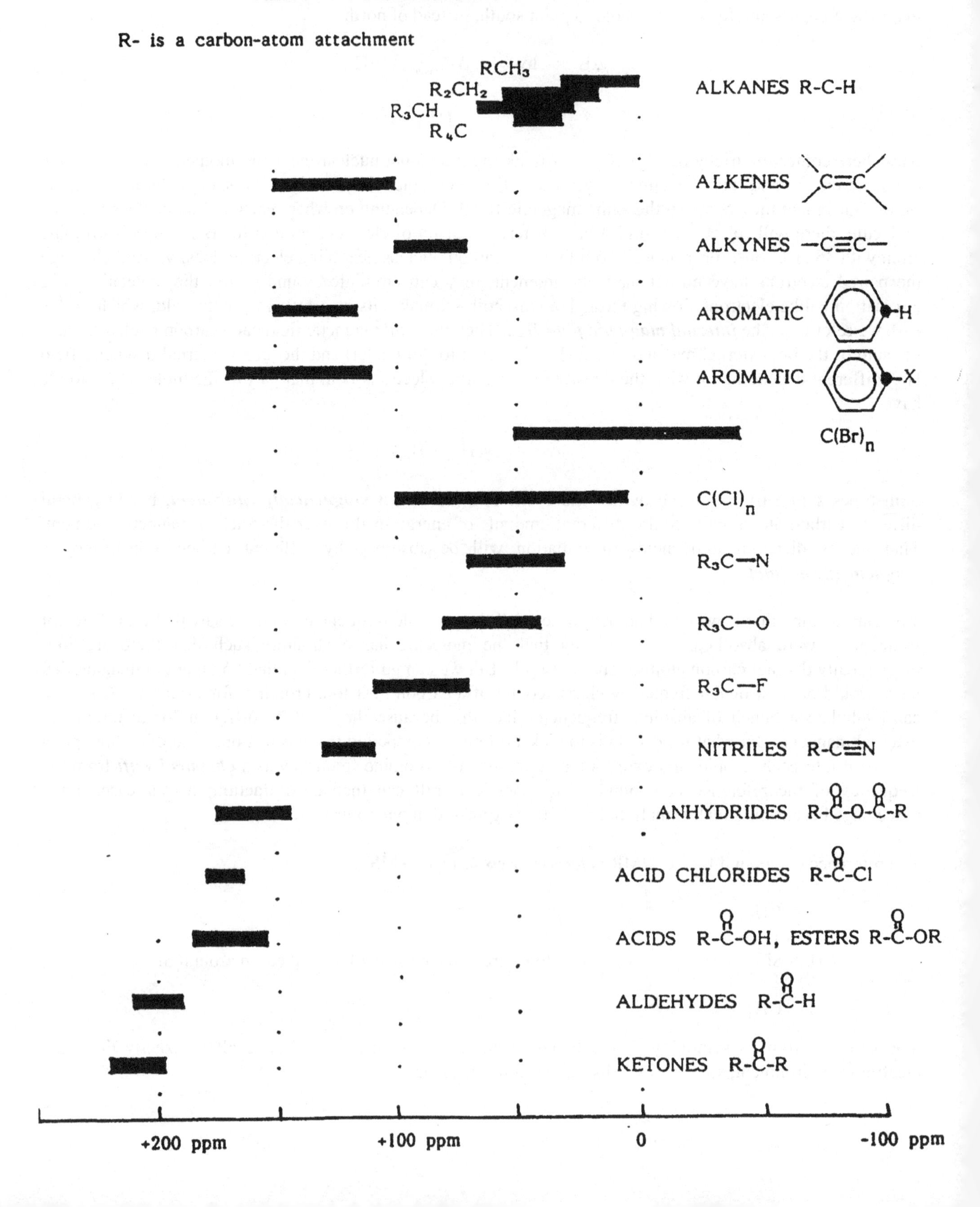

chemical shift in *ppm*, not in frequencies. (You could convert to frequencies, because one part per million of 100 MHz is just 100 Hz, but there's usually no need to.) Note that a carbon next to one or more bromines, for example, will have a chemical shift somewhere between –50 and +50 ppm.

Let's look at a couple of examples of ^{13}C NMR spectra and interpret them according to these principles. Figure 4 shows the ^{13}C NMR spectrum of ***1-chlorobutane***, whose structural formula is given by the spectrum. The chlorine atom will obviously have more electrons than the hydrogens do, and it will have more electron circulation and thus more internal magnetic field. So, given the molecule's structure, there should be four magnetically distinct carbon atoms (one next to the Cl, another one atom away, another two atoms away, and one three atoms away) and thus four absorption frequencies and four peaks on the spectrum. Bingo. The

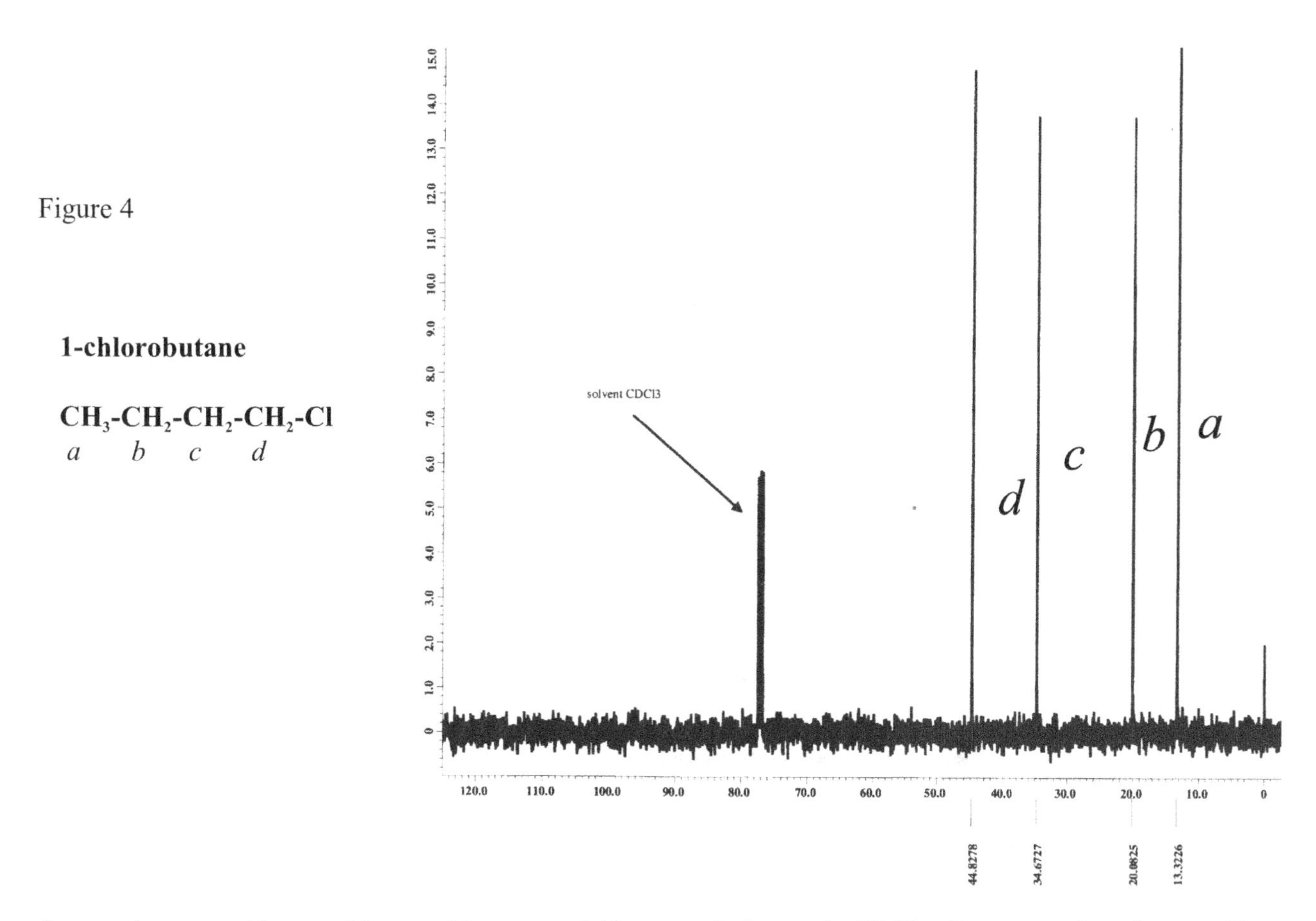

four peaks are at 45 ppm, 35 ppm, 20 ppm, and 13 ppm relative to the TMS reference peak at 0 ppm. These four chemical shift values fit right into the bars on the correlation chart in Table IV for $C(Cl)_n$, R_2CH_2 alkane, and RCH_3 alkane, though it might be a little tough to decide which is which. Note that there's a rubber peach in the fruit bowl in the form of a fat peak (actually three equal peaks side by side, a *triplet*) at 77 ppm; that's due to the carbon atom in the deuterochloroform $HCCl_3$ solvent we used. Since most NMR spectra are run in chloroform solution, that peak will always be there but doesn't mean anything.

Note that we distinguish the magnetically distinct carbon atoms with letters, and those letters will also apply to the peaks on the spectrum; when you get a spectrum, letter the peaks from right to left, *a* to whatever, and use those letters to tie the peaks to a molecular structure as we've done here.

Figure 5 on the next page shows the spectrum of ***acetone***, CH_3COCH_3. The TMS peak is again at the far right end of the spectrum, and there are two other peaks due to the acetone, at 30 and 206 ppm. Now there are three carbon atoms in acetone, not two – but note that the symmetry of the molecule means that the two CH_3 carbons are magnetically equivalent. So the peak at 206 ppm is due to the C=O carbon (see the *ketone* line in Table IV), and the peak at 30 ppm is due to the **two** CH_3 carbons. Notice that there's a whopping chemical shift due to the presence of a double bond to O on the *carbonyl* carbon atom. It's easy to distinguish from other kinds of carbon atoms. Note also that the peak at 206 ppm isn't very tall. You'd think that because it's due to one carbon and the peak at 30 ppm is due to two carbons, the two peaks would be in a height ratio of 2:1, but this is more like 5:1. There's some worthwhile information hiding in here. It turns out that a C with a H bonded to it can flip its nuclear spin faster (relax faster), and that means carbon atoms with *no* hydrogens won't completely relax while we're running the spectrum, and their peak will be smaller. So in general if one peak in a ^{13}C NMR spectrum is suspiciously small, it probably represents a carbon with no hydrogens bonded to it.

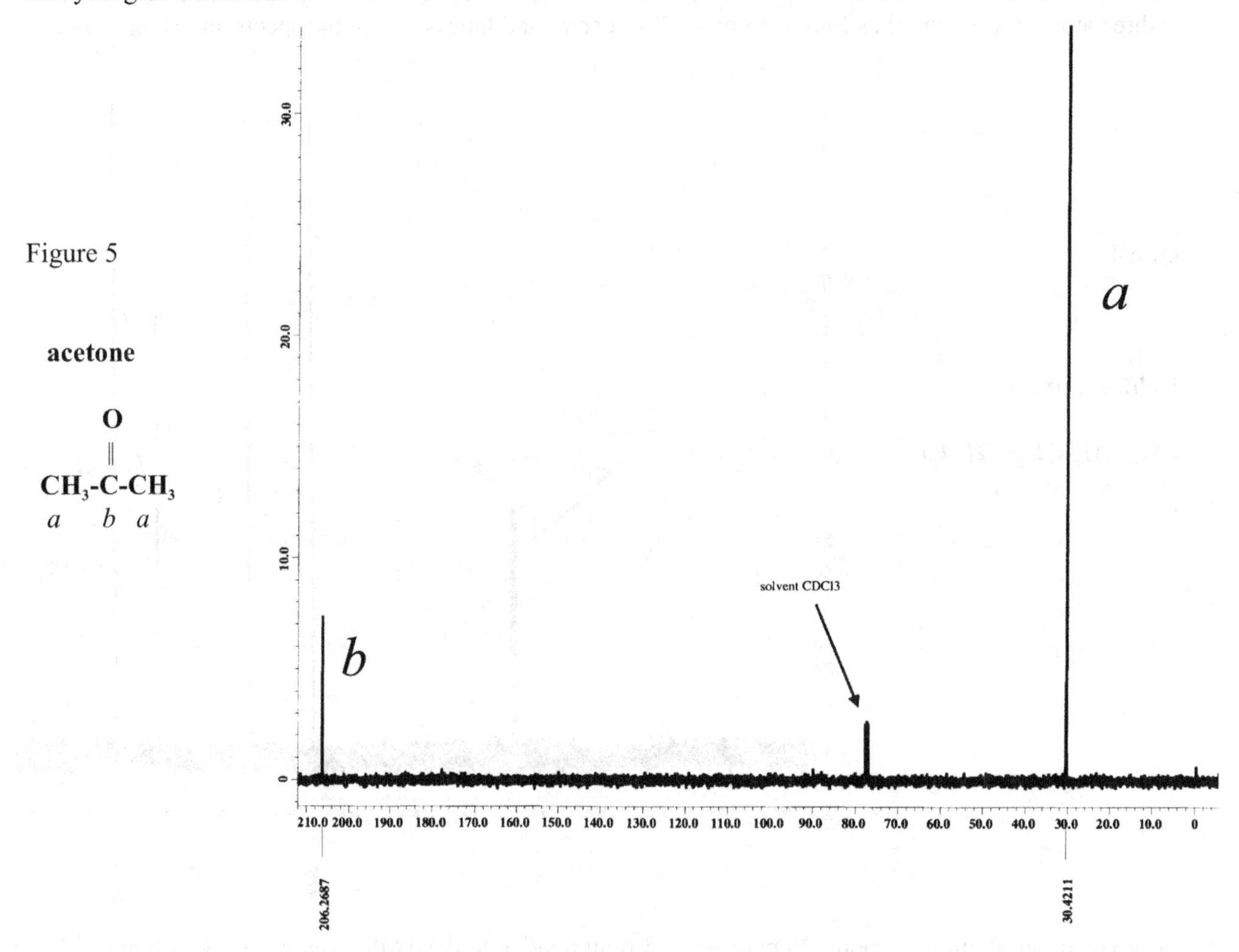

Figure 5

Figure 6, also on the next page, makes the point that symmetry is all-important for NMR spectra by comparing the spectra of benzene, C_6H_6, and phenol, C_6H_5OH. As the structures show, these both have a six-membered ring of carbon atoms (it's understood that there's a C-H at each corner of the hexagon unless something else is shown in place of the -H). But in benzene there's perfect hexagonal symmetry of the nuclei and electrons – all the carbon atoms are in exactly the same electronic environment. So there is only one frequency emitted by the carbon nuclei when they relax, and only one peak shows up in the ^{13}C NMR spectrum.

On the other hand, the six-membered ring in phenol loses its sixfold symmetry, because one of the carbon atoms has an OH on it instead of a H atom. There is still some symmetry left, because the two carbons labeled *a* on the structure are equivalent in that they are both next to a benzene (aromatic) carbon with an OH on it. Similarly, the two carbons labeled *c* are equivalent. But carbon *b* and carbon *d* are unique. We expect to see four peaks, all with chemical shifts in the ranges indicated by the aromatic bars in Table IV. Note the peak heights are roughly proportional to the number of carbons, and that the really little peak *d* is just the carbon with no H.

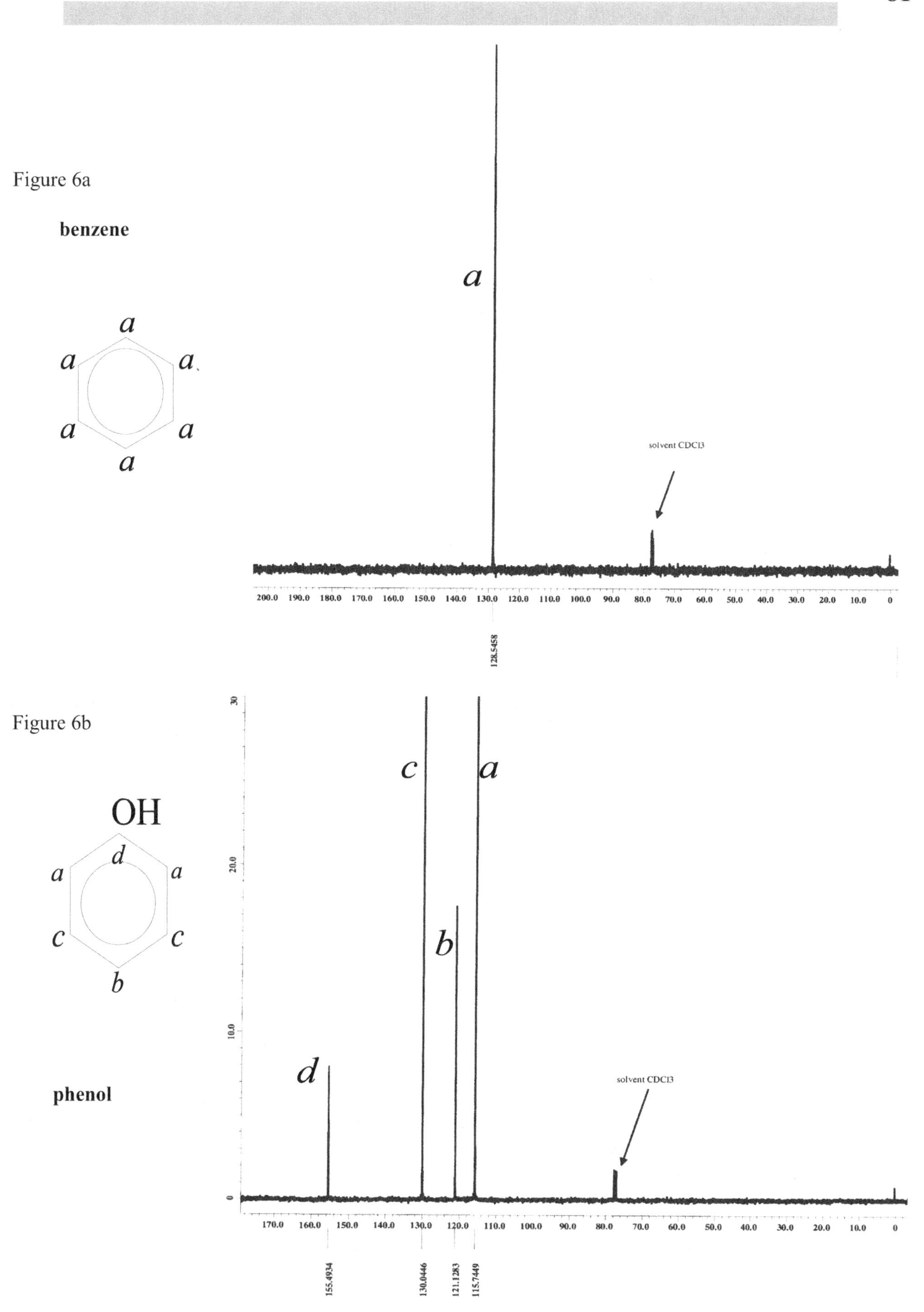
Figure 6a
benzene
a
a
a
a
a
a
a
solvent CDCl3
200.0 190.0 180.0 170.0 160.0 150.0 140.0 130.0 120.0 110.0 100.0 90.0 80.0 70.0 60.0 50.0 40.0 30.0 20.0 10.0 0
128.5458
Figure 6b
OH
d
a
a
c
c
b
phenol
30
20.0
10.0
0
c
a
b
d
solvent CDCl3
170.0 160.0 150.0 140.0 130.0 120.0 110.0 100.0 90.0 80.0 70.0 60.0 50.0 40.0 30.0 20.0 10.0 0
155.4934
130.0446
121.1283
115.7449

Finally, Figure 7 shows what happens to the spectrum of a benzene-ring compound when its symmetry is completely destroyed, as it is in ***o*-chlorophenol**, C_6H_4ClOH. Now the carbon with a Cl (labeled *b*) and the carbon with an OH (*f*) are unique, but so are the carbon next to the C-Cl (*e*) and the one next to the C-OH (*a*), and so are the ones next to them (*c* and *d*). We expect to see six peaks, though they will be closely spaced in the range indicated by the aromatic bars in Table IV. Again, note that the peak heights are very roughly proportional to the number of C atoms of each type (one), and that the two small peaks *b* and *f* are small not because they represent fewer C atoms but because they each have no bonded H atom.

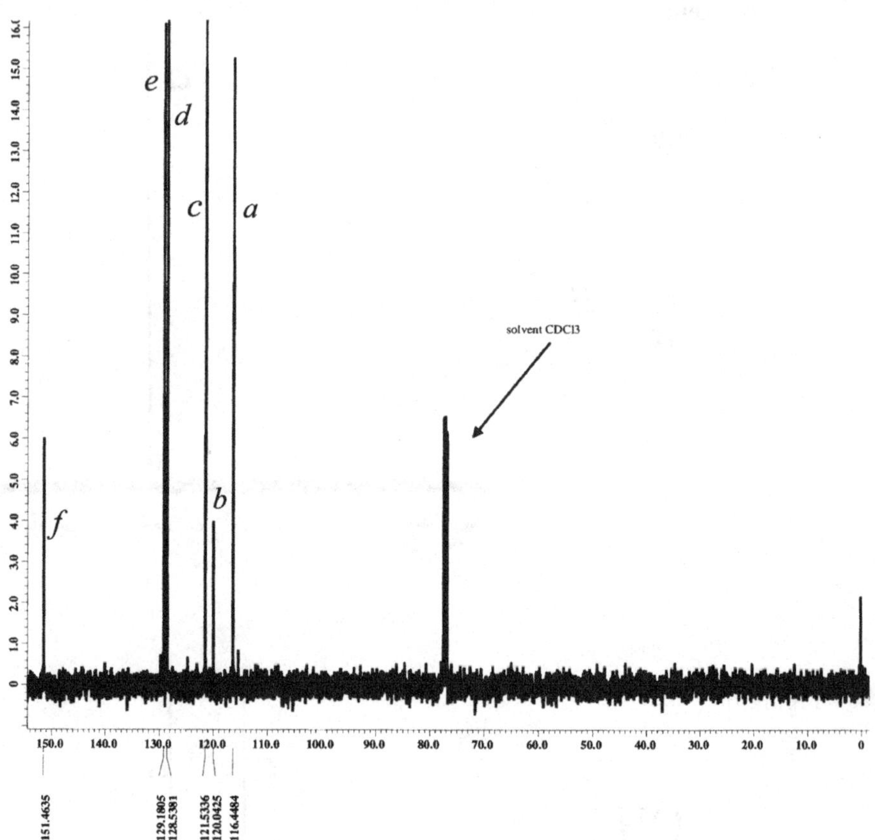

Figure 7

***o*-chlorophenol**

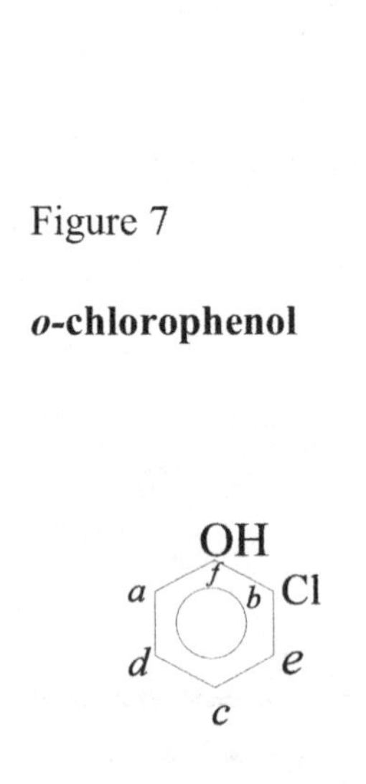

How the NMR Instrument Gives Us Spectra

What we need here is an instrument that can (1) send radiation to our sample to flip all the ^{13}C nuclear magnetic moments, then (2) tell which specific frequencies have been absorbed, and (3) print out the NMR spectrum as a graph of intensity vs. frequency. Look at Figure 8; it's a general flow chart for the signal in the instrument, so follow the arrows.

Obviously we have to have the big constant external magnetic field, so we start with

- the **high-field magnet**, which is a superconducting solenoid cooled by liquid helium. Since it's not connected to the wall socket, it can't be affected by power fluctuations, and it gives a constant magnetic field for years at a time. Now start looking at the signal flow. Start with
- the radio-frequency **RF transmitter**, which generates a *band* of frequencies near 100 MHz, covering all the frequencies a ^{13}C nucleus anywhere in any molecule could possibly absorb. It broadcasts these frequencies into the sample through
- the **transmitting antenna**, which is a coil inside the magnet housing that focuses the radiation on the sample – which is in a narrow tube right in the center of the magnet, spinning rapidly to average out any fluctuations in the magnetic field. When the ^{13}C nuclei in the sample absorb the energy and flip over their nuclear magnetic moments, the RF transmitter turns off, so as not to confuse what happens next. The RF transmitter has produced an ***excitation pulse*** of radiation.

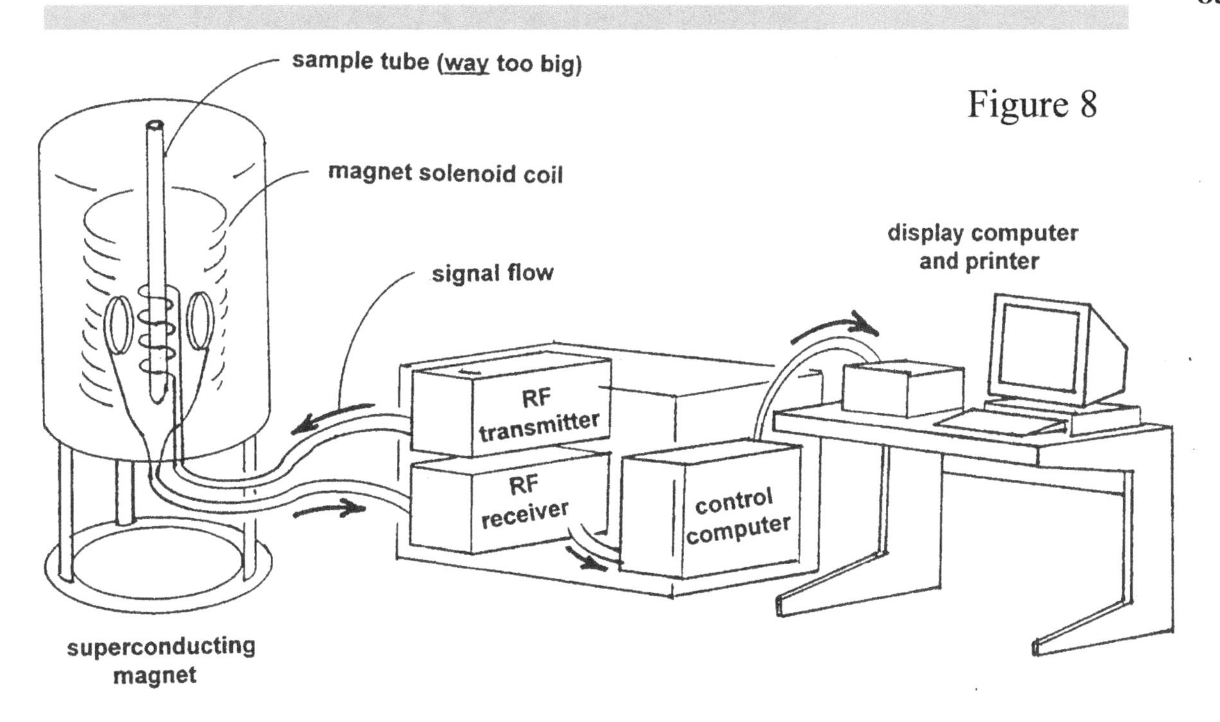

Now what we have is a sample tube full of ^{13}C nuclear magnetic moments, all pointing antiparallel to the external (and internal as an add-on) magnetic field – and no radiation is coming in. So what's going to happen is that the antiparallel magnetic moments will flip back (***relax***) to the parallel, low-energy direction, and they will lose the energy they've just absorbed by re-broadcasting it out again. But they won't all relax at the same instant in time; it takes a while, and the number relaxing at any given instant drops off slowly until they're all back down in the low-energy state. So the sample tube is now broadcasting RF radiation, but not a broad band as the transmitter did. Instead, only *specific* frequencies are being broadcast, one for each carbon-atom environment in the molecule; maybe only one frequency in a really simple case, or maybe four or five or fifteen frequencies. But they're all being broadcast at the same time. The NMR instrument has to be able to sense them, so it has

- the **receiving antenna**, which is another coil around the sample tube, but at right angles to the transmitting antenna. When the excitation pulse has turned off, the receiving antenna picks up the relaxation re-broadcast signal and sends it to
- the **RF receiver**, which is like an FM receiver except it isn't tuned to a single frequency. Instead, it picks up all the frequencies near 100 MHz simultaneously. It's as if your radio was simultaneously giving you rap music, a talk show, and a speech by the Secretary of State – a real mess of interfering signals. We can look at a graph of this mess on the computer display screen. It's called the FID (***free induction decay***), and it's a graph of signal *intensity* (all frequencies together) vs. *time* (because we're going to collect the signal until all the excited nuclei have relaxed).

Now look at Figure 9. This is the FID graph for a signal from the ^{13}C nuclei in methanol (CH_3OH), which only has one carbon atom in the molecule and thus obviously only one electronic environment for C. The graph shows a single frequency, whose intensity starts high but tails off as fewer and fewer excited ^{13}C nuclei are left, until there's no signal left. So we stop collecting it.

But wait! There's also a second graph along the bottom, which is *intensity* vs. *frequency*. We've transformed the FID (the top graph) into a NMR spectrum; this one's simple, because there is only one frequency, so just a single spike on the graph. This is a convenient way to see the number of carbon

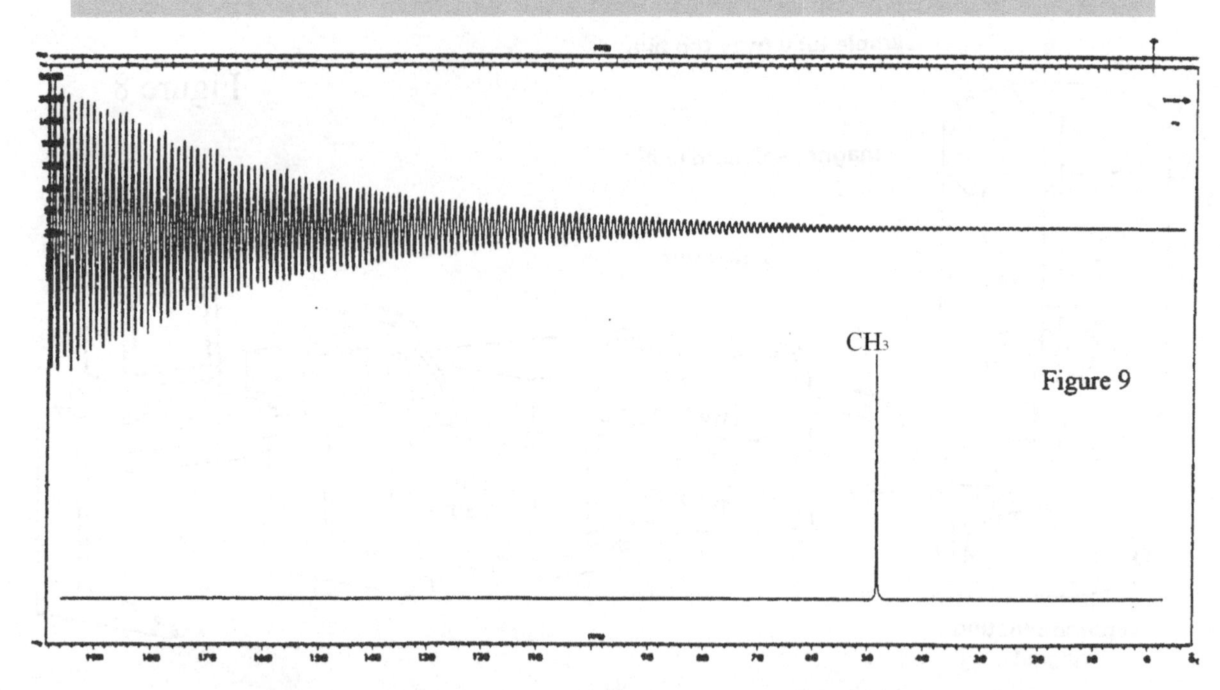

Free induction decay (above) and transformed ^{13}C spectrum of methanol (60% in $CDCl_3$, 25.2 MHz).

environments in the molecule, and even figure out what the environments are, but how did we get this? We used

- the **control computer**, which tells the transmitter when to turn on and off and the receiver when to stop collecting signal. It also digitizes the receiver signal in order to be able to
- perform a ***Fourier transform*** (pronounced Foor-ee-ay) on the digitized FID to give the spectrum.

What's a Fourier transform?

There's an interesting relationship between *time* (the independent variable of the FID graph) and *frequency* (the independent variable of the spectrum). Frequency comes in Hertz units, Hz, but a frequency of one Hertz is just one wave coming by per second. So a Hertz, in fundamental units, is just a per-second, or reciprocal second, sec^{-1}. In other words, frequency is just *reciprocal time.*

What the control computer does is to take the FID graph (*intensity* vs. *time*) and use the mathematical Fourier transform procedure to re-plot the graph as a spectrum (*intensity* vs. reciprocal time, or *frequency*). You can sort of imagine how to do this, because if we gave you a graph of y vs. x and a clean sheet of graph paper and told you to plot y against $1/x$, you could do it; it would just be tedious. The great thing about computers is that they don't mind tedious. So as soon as you've collected the FID a number of times, so as to average out noise, the control computer (which has already let you see what the FID looks like as it's being collected) performs a Fourier transform on the averaged-out FID graph to yield the intensity vs. frequency NMR spectrum. Then it sends the spectrum graph to

- the **display computer**, which displays the whole spectrum on the screen and lets you print it, but also allows you to pick portions of the spectrum to blow up, in case several peaks are really close together.

Now, we've illustrated the process for the really simple FID and spectrum for methanol, but you have to realize that this can get to be a very messy interference pattern for multiple frequencies. Look at Figure 10, which shows a synthetic graph of two frequencies close together – in this case 56 Hz and 64 Hz. If we plot

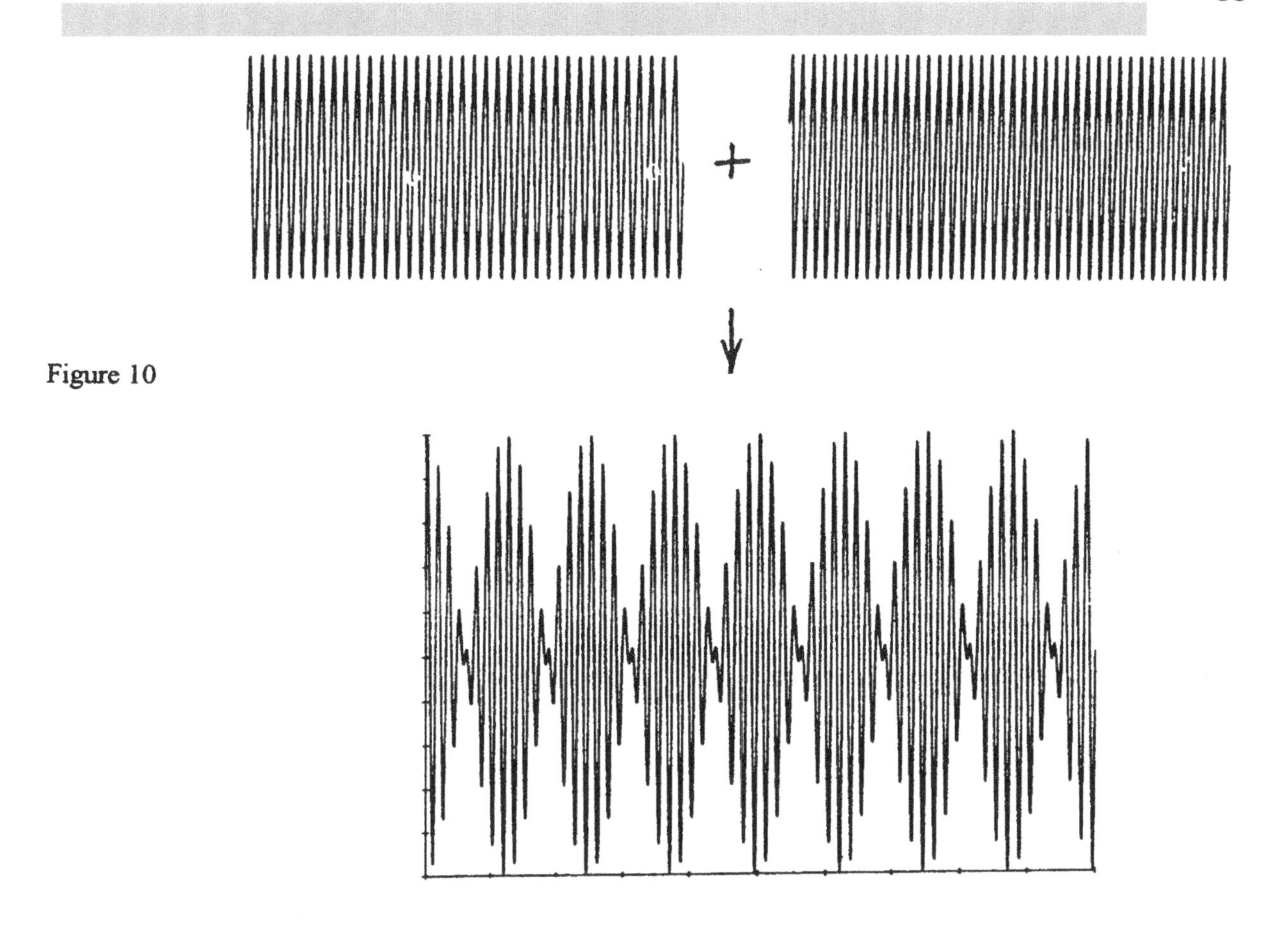

Figure 10

the graph of the two frequencies added together, what happens is that every eight cycles the crest from one wave just cancels the trough from the other wave (***destructive interference***). On the other hand, in between the two waves add together (***constructive interference***) and we get a big signal. So there is a ***beat frequency*** in the combined wave – a long, low-frequency wave on top of the high-frequency wave.

This is what's going on in Figure 11, the FID (and spectrum) for ethanol, CH_3CH_2OH, which has two different carbon environments and thus re-broadcasts two frequencies. The frequencies are interfering with each other, and you can see the long, slow beat frequency.

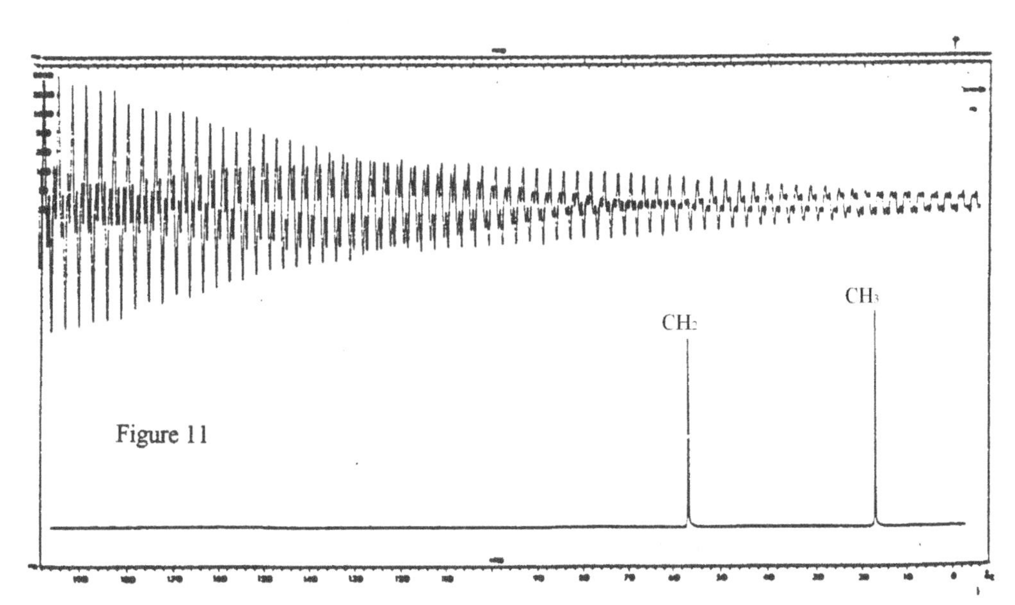

Figure 11

Free induction decay (above) and transformed ^{13}C spectrum of ethanol (60% in $CDCl_3$, 25.2 MHz).

In Figure 12 you can see what happens when a Fourier transform is applied to the synthetic graph of two frequencies: it just produces two spikes at 56 Hz and 64 Hz, which were of course the original frequencies. And that's how the two spikes in the spectrum at the bottom of Figure 11 were obtained.

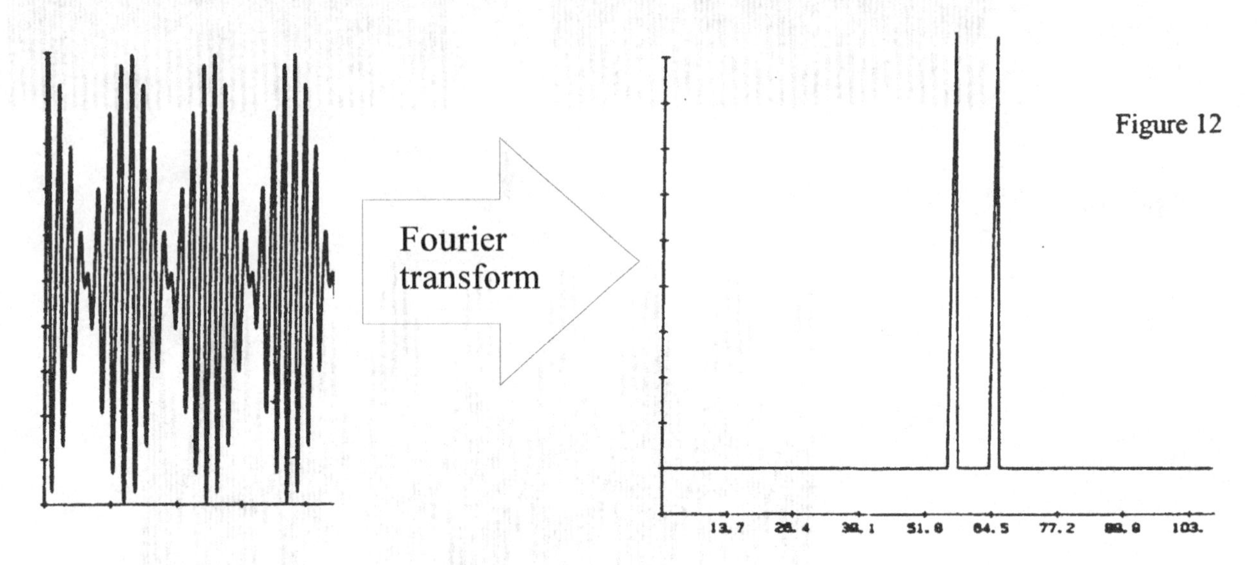

Figure 12

The FT-NMR Technique Quiz

Study the above discussion and your lab lecture notes till you feel okay about (1) what's happening in the molecule when ^{13}C NMR is being done to it, and (2) how the instrument shows us what frequencies are absorbed by the molecule and thus what C environments are present in the molecule.

Then sit down with your list of LB candidate compounds and pick out the three closest to the *bottom* temperature you observed in the distillation. For each compound, make out a little table like the one below on a right-hand page in your notebook (but obviously not for this compound unless it's one of your candidates). Use the correlation chart (Table IV) to figure out the ranges, and note that you should look for the most unusual thing about a C atom's environment to figure out which range to use:

a *b* *c* *d*
CH_3 - CH_2 - CH_2 - CH_2 - Br

C atom	C Type and Chem Shift Range	
a	Alkane RCH_3	0 – 30 ppm
b	Alkane R_2CH_2	20 – 55
c	Alkane R_2CH_2	20 – 55
d	$C(Br)_n$	-50 – 50

When you get these three tables done, pick out the three HB candidates closest to the *top* of the boiling point range you observed. For each compound, make out a similar table, using the correlation chart to pick the chemical shift ranges. When you get through, you'll have predicted six ^{13}C NMR spectra.

Now find a faculty member and ask to take the quiz on ^{13}C NMR. The quiz consists of the following questions:

(1) What happens inside a molecule when the ^{13}C experiment is being performed?

This is the material back on pages 76-77. For this question, you don't need to say anything about the instrument hardware except to note that a big magnet is needed to align the magnetic moments in the first place.

(2) How does the FT-NMR instrument operate to tell us which frequencies of radiation are characteristic of our sample?

This asks for a guided tour of the instrument in the sequence given by the bulleted list above (pages 82-83) and including a short version of how the Fourier transform process works.

These first two questions you'll need to be able to answer out of your head, with no notes. Then you'll need your notebook for the third question:

(3) What predicted spectra have you written up for your three best LB candidate compounds and your three best HB candidate compounds?

For this question, give us your notebook; we'll go through your predictions and help you clean them up if there seem to be any mistakes present.

Once you've satisfactorily answered these three questions (and written up the sample prep procedure), the faculty member will initial your notebook and you can sign up for instrument time. **To receive credit for the ^{13}C NMR experiment, your notebook *must* be initialed; just having spectra there is not enough!**

Sample Preparation (about 20 minutes)
The NMR sample tubes are very thin-walled and easy to break, and surprisingly expensive (about $5), so be careful! You should have checked out two sample tubes and caps, one yellow and one blue, already. Choose the vials with the purest low-boiling and high-boiling liquids and take the vials, sample tubes, and two *clean dry* micropipet eyedroppers to the NMR sample-filling block in the NMR lab. Place a sample tube in the slot at the front; it should go down roughly two inches. Use one micropipet to fill the tube up to the top of the yellow paint, about halfway to the top of the wood block, with your *low-boiling* compound. Then use the automatic filler on the nearby bottle of ***deuterochloroform*** to deliver one squirt of that solvent into the sample tube, which should get the liquid level to the top of the wood block. Cap the sample tube with the *yellow* cap (so yellow = low-boiling), and invert it 5 times. Check to see that the sample is completely mixed and clear; if it's still separated, or cloudy, see the instructor.

Put the other sample tube in the block and fill it similarly, using the other micropipet to avoid contamination, with the *high-boiling* compound, add deuterochloroform, check the level, cap it with the *blue* cap (blue = high-boiling), and invert it 5 times. Check for complete solution. Write the two color codes down in your notebook. Wipe off the capped tubes with a Kimwipe to make sure they're absolutely clean and dry outside. Take the sample tubes with you when you go to your instrument appointment.

Running the spectrum and saving the sample
The instrument technician will insert the sample tube in the magnet for you and log you into the computer for the instrument. He'll walk you through the menu on the screen, but ***you*** should use the mouse to make choices with his help. Once you get a complete spectrum on the screen, it's possible you'll have two peaks so close that it's hard to distinguish them; if this happens, you can draw a box around the two and blow that section up to see it clearly as a separate little box out on an empty part of the spectrum. When it looks right on the screen, print it. The printed spectrum will be red, with the grid and numbers in black.

When you get your printed spectrum, the chemical shift values will be printed below each peak; ignore the decimals and just think about integer values. In your notebook, you're at the ***Data*** stage, so put that heading at the top of a new notebook page. Mount the spectrum for your low-boiling compound with clear tape on the ***left-hand page***, right side up, folding it once in the middle so it will fit. Interpret on the right-hand page.

Take the two color-coded sample tubes (still full) back to your lab desk and put the two tubes *upright* in an Erlenmeyer flask inside your drawer or cabinet to same them for a later NMR experiment. Make sure they're tightly capped, of course.

Interpreting the spectrum and writing it up

There's a sequence to go through in doing the interpretation (which needs to be written up on the ***right-hand page*** opposite the spectrum).

- First, count the peaks that are present. Ignore a small triple peak at 77 ppm – that's just the chloroform solvent you used. Also watch out for a small peak at 31 ppm that *might* be from acetone the tube was rinsed out with, but hasn't evaporated – but it might be real too; don't throw it away immediately. Usually it's easy to decide how many real peaks you have, but sometimes there are small peaks that might just be impurities – or they might come from a small amount of your other unknown still present. Consult with us if you have trouble deciding.

- Then go back to your list of LB candidate compounds with the right boiling point. Go through and pick out the ones with the *right number of carbons*. On your right-hand data page here, create a *new* short list of compounds that have the right BP ***and*** the right number of carbons. There'll probably be only three or four.

- Now, on your spectrum, letter each peak (that you're counting) with a CAPITAL LETTER: A, B, C, etc., from left to right. The caps are so you won't mix your experimental spectrum up with your predictions.

- Start at the left side of your spectrum – the largest chemical shift value – at peak A. Go back to the correlation chart (Table IV). Run your finger up the chart from the chemical shift for peak A. What kind of functional-group carbon atom fits the chemical shift? (There might be more than one.)

- Below your short list, start your interpretation by describing peak A: the kind of group it might be in or be next to. If it pretty well specifies a functional group (for instance, a peak at 204 ppm almost guarantees either an aldehyde or a ketone), go back through your short list at the top of the page and rule out the molecules that *don't* have that functional group. On the other hand, if there's no peak above, say, 150 ppm, you can be sure you don't have the functional groups that would give higher chemical shifts. Either way, discuss briefly what peak A is telling you.

- Move on to peak B and do the same thing as for peak A. Then peak C, and so on. Note that sometimes a group of peaks might be related to each other: four peaks between about 110 and 160 ppm are likely to be the four magnetically distinct carbons in a benzene ring with one substituent on it(as in Figure 6b), and three peaks between about 10 and 35 ppm might well be three hydrocarbon-chain type carbon atoms as in a propyl group. Include any groups like this in your discussion – but note that this isn't a massive writing exercise; about half a page should cover the whole thing. Finish up with your take on what the possible LB compounds are at this stage.

- Go on to your spectrum for the high-boiling compound. Go to a new blank page in your notebook and do exactly the same series of things you just did for the LB compound: spectrum on the left, interpretation on the right. Again, get the number of carbons, get a new short list, interpret the experimental peaks from left to right, and finish up with the remaining HB possibilities.

- When you've interpreted both spectra, enter the heading ***Summary***, and write up a very short bit on where you are as a result of having both BP data and ^{13}C NMR data.

Now, let's take Irv Gratch through his high-boiling unknown to see how this works out. Irv's BP range for his HB liquid was 122-124°C. Allowing for a band around these temperatures, he sieved the table and got about 15 possible compounds, but here we can just look at five that are a good BP fit:

Butyl acetate	$CH_3COO(CH_2)_3CH_3$	124	6
2,4-Dimethyl-3-pentanone	$[(CH_3)_2CH]_2CO$	124	3
Ethyl butyrate	$CH_3(CH_2)_2COOCH_2CH_3$	120	6
3-Hexanone	$CH_3CH_2CO(CH_2)_2CH_3$	123	6
3-Methyl-3-pentanol	$(CH_3CH_2)_2C(CH_3)OH$	123	4

So Irv has passed the quiz, made up samples, and gotten printouts of the spectra. Here's the spectrum for his high-boiling component:

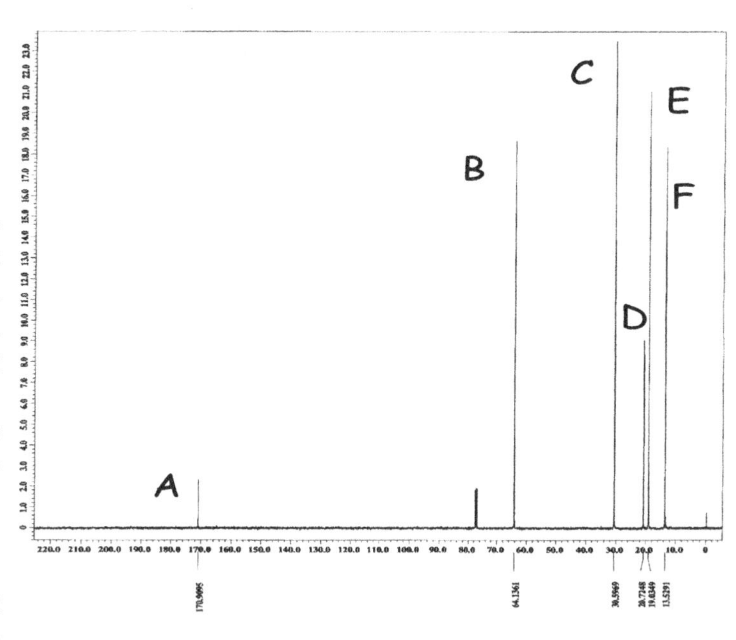

First, he counts the peaks: The three little ones at 77 ppm are clearly from the deuterochloroform solvent, so they don't count. Then there's one at about 31, which could possibly an acetone impurity – but it's a big peak, and probably it's real. Assuming it's OK, he counts 6 peaks, at 171, 64, 31, 21, 19, and 14 ppm. On this assumption, he letters the peaks A-F, as shown on the spectrum.

In the next step he looks at his boiling point possibilities (here, just the five shown above) and pick out the ones with the *right number* of peaks – the "sixies". That gets him three possibilities, and he makes out a new short list of those three:

Butyl acetate	$CH_3COO(CH_2)_3CH_3$	124	6
Ethyl butyrate	$CH_3(CH_2)_2COOCH_2CH_3$	120	6
3-Hexanone	$CH_3CH_2CO(CH_2)_2CH_3$	123	6

Now, he interprets the *chemical shifts* of the six peaks, one at a time, starting at the left with peak A at 171 ppm. Using the correlation chart on page 80, he runs his finger up from the bottom at about 170 ppm to see what functional groups absorb about there. He quickly notes that neither aldehydes nor ketones fit – so he can be sure he *doesn't* have an aldehyde or ketone. But what *does* he have? Acids, esters, acid chlorides, and anhydrides all seem to absorb about 170 ppm. Further up the chart, a substituted benzene-ring carbon (aromatic C-X) absorbs there, but there's only one peak. If it were a substituted benzene, there would have to be some nearby peaks due to the *other* benzene-ring carbons. He looks closely at the chemical-bond structures to see what functional groups his three short-list candidates have. The first two are *esters* (R-CO-OR), while the third is a *ketone* (R-CO-R). But ketones are out – so now he's down to two possibilities. He writes:

> A - 171 ppm matches an ester C=O (candidates 1 and 2) but not a ketone (candidate 3), which would require a peak around 200-220 ppm.

Now he goes on to peak B; it's at 64 ppm, which from the correlation chart is C-F, C-O, C-N, C-Cl, or just possibly a R_3CH *alkane*. Esters don't have F, N, or Cl, but they do have C-O type carbons – so maybe that's the peak at 64 ppm:

> B - 64 ppm matches an ester C-O; if one of the two possible esters here doesn't work out, the original BP chart will have to be sieved again for C-F, another C-O, C-N, or C-Cl (but almost certainly an ester).

Now the peaks at 31, 21, 19, and 14 ppm (peaks C, D, E, and F) look like a hydrocarbon cluster; they all fall in the range at the top of the correlation chart for *alkanes*, and both the esters under consideration have four carbons that aren't either C-O or C=O, just C-C:

> C, D, E, F - 31-14 ppm all match alkane CH_3 or CH_2 ranges, which are seen in both molecular structures being considered.

At this point, he has narrowed his list of candidate compounds down to two *and* has fully interpreted his spectrum in terms of both the number of peaks and the meaning of each chemical shift. It's time to go on to infrared spectra – of course, he'll use these results to make IR predictions.

But just for the record, the opposite page shows what Irv came up with in his notebook.

25

Irv Gratch
2/4/89

Experiment 3: ^{13}C NMR on High-boiling Compound

Purpose: To establish number of magnetically distinct C sites in HB compound and evaluate possible functional groups

Procedure:

1) Predict spectra for leading candidates (BP 118–126 °C)
2) Pass oral quiz on FT-NMR fundamentals
3) Fill NMR sample tube: sample HB cpd to yellow band fill block, $CDCl_3$ to top of block, invert 3x
4) Have instrument tech mount sample tube in instrument
5) Run "1D 13C" spectrum

Apparatus: JEOL Eclipse 400+ FT-NMR spectrometer, 5-mm sample tube, filling-template block

Data: Blue cap = HB sample

Spectrum at left → 6 peaks A–F:

A: 171 ppm matches an ester C=O (candidates 1 + 2) but not a ketone (candidate 3), which would require a peak 200–220 ppm

B: 64 ppm matches an ester C–O; if one of the 2 possible esters here doesn't work out, the original BP chart will have to be sieved again for C–F, C–O, C–N, or C–Cl (but almost certainly an ester).

C, D, E, F: 31–14 ppm all match alkane CH_3 or CH_2 ranges, seen in both molecular structures being considered.

INFRARED SPECTROSCOPY (IR) (after quiz, about 20 min; start on notebook p. 35)

This technique will give you *some* information, sometimes, about the arrangement of carbon atoms in your compound's carbon skeleton, but primarily it gives you direct information about the presence of ***functional groups*** involving heteroatoms. In this case, the radiation that we're interested in has a frequency range of about 10^{13}-10^{14} Hz, slightly lower frequency than visible light. Again we're going to supply energy to the molecule as radiation, but this time it's not possible to observe the molecule re-radiating the energy directly. Instead, what we will do is to supply a constant band of radiation and look at how much of the radiation has been absorbed at each frequency after it passes through the sample.

What does radiation at this frequency do to the molecule? First, it's necessary to appreciate some features of the bonding in organic molecules. The electrons that form each bond (an electron pair for a normal single bond, two pairs for a double bond, etc.) are attracted to both positively-charged nuclei at the ends of the bond, so they hold the atoms together almost like a spring attached to an atom at each end. But the nuclei have almost all the mass in each atom, so it's rather like two wooden balls held together by a very lightweight spring. Under the influence of thermal energy (the molecules in the liquid sample tumbling around and hitting each other), these ball-and-spring combinations will vibrate, stretching out and then compressing back in again. A molecule with only two atoms has only this stretching mode of vibration, but one with three atoms could also have a bending mode in which the natural bond angle opened and closed in another kind of vibration. In general, large complicated molecules will have lots of possible modes of vibration, so that they flop around rhythmically all over the place.

What governs the frequencies of these natural vibrations of the molecule? They're like any other vibration, such as that of a screen-door spring. The vibration will occur faster the stronger the spring is, and it will occur faster the lighter the atoms are (not so much mass for the spring to haul around). So we could expect that a C-H bond would vibrate at a higher frequency than a C-C bond, because the hydrogen atom is so light—and a C=O bond will vibrate at a higher frequency than a C-O bond because the double bond (the equivalent of the spring) is stronger than the single bond. This means that if we can identify the natural vibrational frequencies of a molecule we have a guide to the masses of the atoms and the strengths of the bonds in the molecule. Obviously we ought to be able to correlate this with the presence of particular patterns of chemical bonds in the molecule, such as H-N-H, C=C, or S=O. Atoms other than C and H are usually called ***heteroatoms***, and the specific bonding patterns involving them are known as ***functional groups***.

Now suppose that the identities of the atoms in the bond are different: maybe a C-H bond, or a C=O bond. The different nuclei have different degrees of attraction for the bonding electrons, and thus different degrees of ownership of the electrons. This means that the electrons will be unequally divided between the two atoms, and there will be a partial positive charge on one atom and a partial negative charge on the other. Such a bond forms an electric dipole: positive on one end and negative on the other.

Here comes the electromagnetic radiation toward the molecule and the polar bond we've just described. The electromagnetic radiation has an electric field associated with it that's positive on one end and negative on the other, and that polarity alternates with a frequency that is the frequency of the radiation. Suppose that at a given moment the radiation electric field is positive on the top and negative on the bottom, and the dipole of the bond is negative on the top and positive on the bottom, as in Figure 13a on the next page. When that's true the electric field of the radiation will pull the atoms away from each other and stretch the bond. But then the natural frequency of the radiation changes the polarity of the electric field, and it's negative on the top and positive on the bottom as in Figure 13b. The field will now repel the charged atoms toward each other, compressing the bond. If this rhythmic stretching and compression of the bond by the field of the radiation occurs at the same frequency that the bond naturally *wants* to vibrate (stretch and compress) naturally, the bond vibration will absorb energy from the radiation and will have a greater amplitude of motion of the atoms. If the radiation frequency is different from that of the bond vibration, the two patterns of induced motion will just fight each other and cancel out over a period of time, with no net energy absorption. So the only frequencies of radiation that will be absorbed are those that equal the natural vibrational frequencies of the bonds in the molecule. We can see this in an appropriate instrument, because the frequencies of radiation that have had energy absorbed out of them will be less intense—dimmer—when they pass out of the sample, while those that have not been absorbed will have all their original intensity. And knowing the radiation frequencies that are absorbed is the same thing as knowing the vibrational

Figure 13

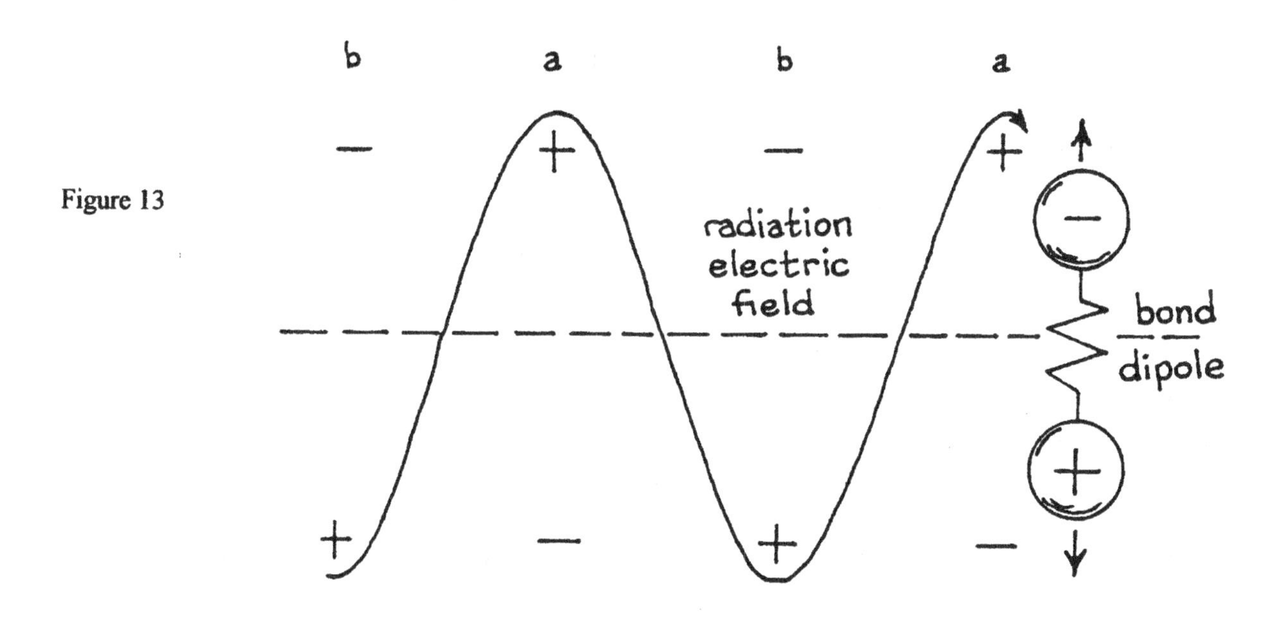

frequencies of the molecule. We use the infrared region of the spectrum because it covers the vibrational frequencies we're interested in.

Chemists have spent several decades accumulating information about the vibrational frequencies of carbon skeletons and functional-group bond sequences. That information is accumulated in a correlation chart given on the next two pages as Table V and in tabular form. On the correlation chart there are short bars indicating frequency ranges at which particular groups vibrate. The numbers that appear are really what's called ***wavenumber***, rather than frequency, but it's proportional to frequency:

wavenumber = frequency ÷ speed of light

Chemists generally use wavenumber rather than frequency because the numbers are smaller—it gets rid of all those factors of 10^{13}. Wavenumber comes in ***cm***$^{-1}$ (reciprocal centimeters) if the speed of light is expressed in cm/sec (check out the units yourself), and the "frequencies" across the bottom of the chart are really wavenumbers in cm^{-1}.

The way to use a correlation chart to predict the IR spectrum of a given molecular structure is to recognize first that there will generally be two different kinds of vibrations: those due to the carbon skeleton (on the chart at the top as *ALKANE, ALKENE, ALKYNE*, and *AROMATIC* headings on the left), and those due to the various heteroatom functional groups (headings below the carbon-skeleton headings—*ESTER, ALCOHOL*, etc.). Draw out a molecular structure with all the bonds shown, and make sure you know what kind of functional groups are present (ask us for help with that). Then list first the vibrational frequencies due to the carbon skeleton (for instance, maybe an ethyl group under the *ALKANE* heading) and then those due to functional groups (maybe an aliphatic ketone group under the *KETONE* heading. Together, these would represent the vibrational frequencies of diethyl ketone, or 3-pentanone, $CH_3CH_2COCH_2CH_3$.

Key Frequency (Wavenumber) Ranges for Stretching Vibrations in Organic Functional Groups

Functional Group	Bond that is stretching	Wavenumber Range (cm^{-1})	Characteristic	R =
Alcohol	R–O–H	3300-3400	Broad & Strong	alkyl
Carboxylic acid	R–C(=O)–O–H	2500-3500	Very Broad	H or alkyl
Amine	R_2N–H	3280-3380		H or alkyl
Alkyne	RC≡C–H	3250-3350	Sharp	H or alkyl
Alkene & Aromatic	R_2C=C(R)–H	3000-3100		H or alkyl
Alkane	R_3C–C(R_2)–H	2850-2985		H or alkyl
Nitriles	R–C≡N	2240-2260	Sharp	H or alkyl
Aldehyde*	R–C(=O)–H	1720-1740	Strong	H or alkyl
Ketone*	R–C(=O)–R	1705-1725	Strong	alkyl
Ester*	R–C(=O)–OR	1735-1750	Strong	alkyl
Carboxylic acid*	R–C(=O)–OH	1700-1730	Strong	alkyl
Nitroalkanes	R–N^+(=O)–O^-	1500-1550 1330-1390	*symmetric* *asymmetric*	alkyl
Ethers	R_3C–O–CR_3	1070-1140		H or alkyl
Esters	R–C(=O)–O–CR_3	1160-1210 1030-1100		H or alkyl
Halocarbons	R–Cl R–Br R–I	600-800 550-650 500-700		alkyl
*Aromatic Carbonyls	Ph–C(=O)–Z	Subtract 20 to 40 from value above		

Table V — Correlation Chart for IR Peak Wavenumber in *cm*$^{-1}$

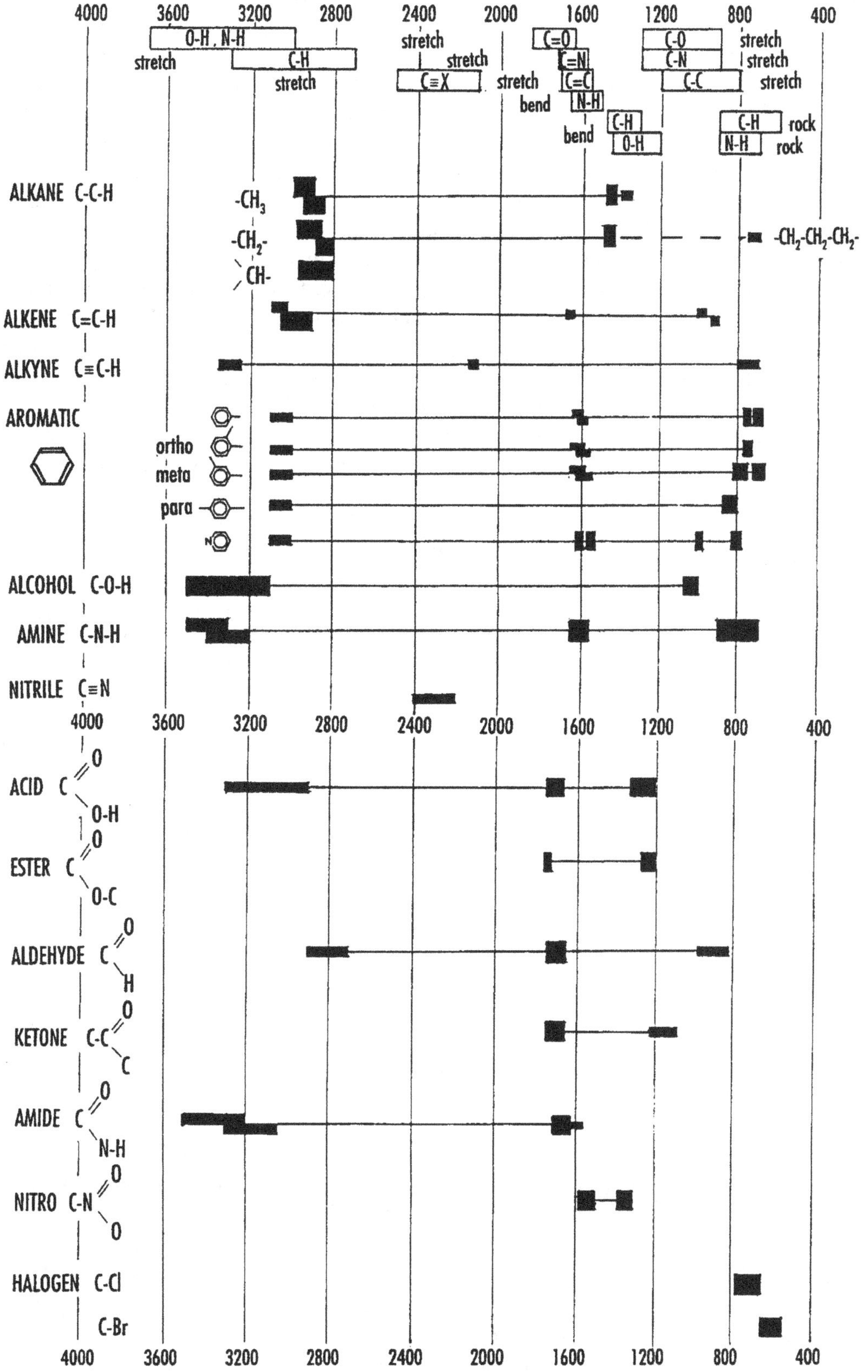

4000 3600 3200 2800 2400 2000 1600 1200 800 400

O-H, N-H stretch
C-H stretch
C≡X stretch
C=O stretch
C=N stretch
C=C stretch
N-H bend
C-H bend
O-H bend
C-O stretch
C-N stretch
C-C stretch
C-H rock
N-H rock

ALKANE C-C-H
$-CH_3$
$-CH_2-$
CH-
$-CH_2-CH_2-CH_2-$

ALKENE C=C-H

ALKYNE C≡C-H

AROMATIC
ortho
meta
para
N

ALCOHOL C-O-H

AMINE C-N-H

NITRILE C≡N

4000 3600 3200 2800 2400 2000 1600 1200 800 400

ACID C=O, O-H

ESTER C=O, O-C

ALDEHYDE C=O, H

KETONE C-C=O, C

AMIDE C=O, N-H

NITRO C-N=O, O

HALOGEN C-Cl

C-Br

4000 3600 3200 2800 2400 2000 1600 1200 800 400

The thickness of the bars on the chart tells you how intense the absorption of radiation should be. What you need to list are the strong absorptions (fat bar), and maybe the medium-strength absorptions (thin bar) if they occur in a really unusual region of the spectrum. The carbon skeleton should give you a couple of absorptions, and the functional groups a couple more, perhaps a total of 3-5 predicted frequencies. Then, with these narrow frequency ranges listed, go up to the very top of the page (open bars with *stretch, bend,* or *rock* printed beside them and the specific bond inside) and, for each predicted absorption, decide what kind of molecular vibration is responsible. On the other hand, you can do some structure prediction from an IR spectrum for an unknown molecule. But IR spectra show lots of complicated kinds of molecular vibrations, and there will be more peaks than you can explain. (In the same sense, an experimental spectrum that you compare with a predicted spectrum always has more peaks than you predicted.) In general, you can predict, or explain, peaks at higher frequencies (wavenumbers) than about 1500 cm^{-1}. Below 1500 cm^{-1}, most of the peaks are fairly complicated combinations of three or four or five bonds all twisting around together, and they're uniquely typical of that one molecular structure, rather than a general category. It's like a fingerprint of the molecule in that sense, and the low-frequency region is often called the "***fingerprint region***". However, there are a few functional group frequencies below 1500 cm^{-1} that are important: ethers (in the table but not on the chart) at about 1100-1200 cm^{-1}, esters at about 1200 cm^{-1} (along with two other peaks near 1700 and 1100, as in the table), Cl and Br at about 600-700 cm^{-1}, and nitro groups at 1500 and 1400 cm^{-1}. If you have a compound with various groups substituted on a benzene ring, the range from about 650-900 cm^{-1} will show where the substituted groups are on the ring. You should be alert to these, and to all the absorptions above 1500 cm^{-1}.

Let's take one example. Figure 14 shows the IR spectrum of 3-pentanone, the compound we mentioned above The first thing to notice is that we're looking at radiation absorbed, not emitted, and so the plot shows absorptions as dips rather than peaks (but chemists still call them "peaks" anyway). Above 1500 cm^{-1}

Figure 14

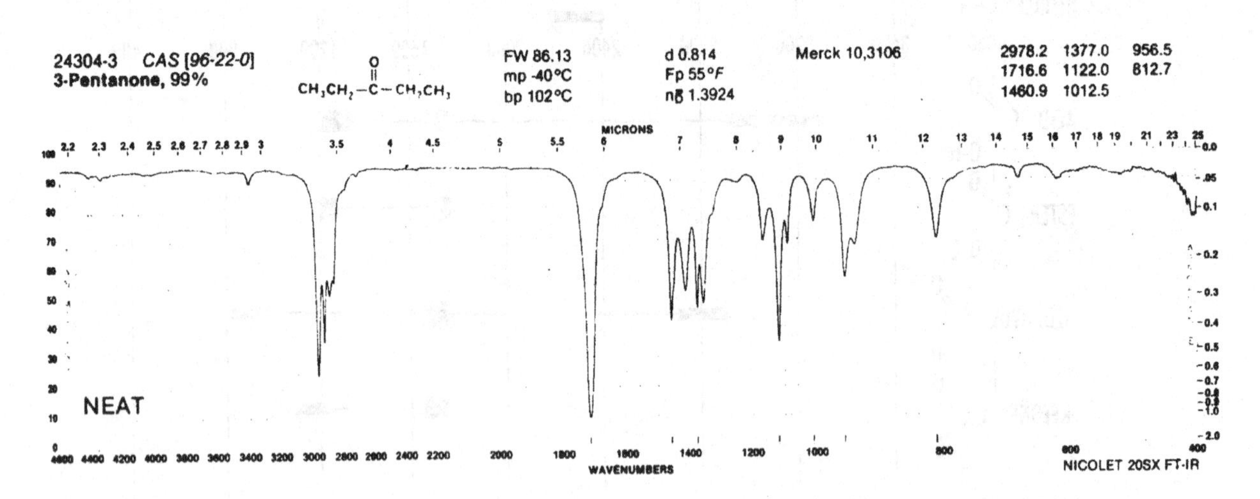

(on the left half of the spectrum) there are only two peaks, which the peak listing in the upper right corner tells us are at 2978.2 cm^{-1} and 1716.6 cm^{-1}. The 2978 peak, however, has a couple of little brothers at slightly lower frequencies, perhaps 2900 and 2930 cm^{-1}. From the top of the correlation chart (CH_3 and CH_2 under *ALKANE*) we can see that there should be a group of three peaks between about 2800 and 3000 cm^{-1} for an ethyl (CH_3CH_2-) group, and that's just what we have. The other high-frequency peak, at 1717 cm^{-1}, is a good fit for the aliphatic ketone under *KETONE*. The intensities are right, too: the ethyl group of peaks should be strong, and the ketone peak should be strong. To see what kind of molecular vibration is involved, we trace the chart bars up to the top of the chart. The 2978 cm^{-1} group are clearly all C-H stretch vibrations, and the 1717 cm^{-1} peak is a C=O stretch vibration. There are about 11 other peaks visible on the spectrum (below 1500 cm^{-1}), but neither the ethyl groups nor the ketone group have important diagnostic absorption peaks there—so, basically, we ignore them. If you wanted to, you could key a number of them to bars shown on the chart for either ethyl or ketone groups, but they're not really important.

In the spectrum of Figure 14, notice some very important high-frequency peaks that are *not* present. There's no O-H or N-H stretch (3700-3000 cm^{-1}), so we can't have an alcohol (O-H) or amine (N-H) functional group in our compound. There's no C≡N peak (2400-2200 cm^{-1}), so we can't have a nitrile group. The C-H peaks below 2978 cm^{-1} are at too low a frequency to be due to an aromatic (benzene-ring) type of compound (3130-3000 cm^{-1}). And the 1717 cm^{-1} peak is probably at too low a frequency to be due to an ester group (1750-1720 cm^{-1}). So besides verifying the 3-pentanone structure that we might have guessed at from the boiling point and ^{13}C NMR, we have *ruled out* several other important categories of compounds.

Fourier-Transform Infrared Instrumentation (FT-IR)

The discussion so far has made it clear that specific identification of vibrational frequencies can give us a lot of information about the structure of a molecule, particularly with respect to the heteroatom functional groups that may be present. This is like the ^{13}C NMR situation, in which specific relaxation frequencies can give us information about electron densities along a chain of carbon atoms. We'd like to have an instrument that does the same thing the FT-NMR does—to look at infrared intensity as a function of time and Fourier-transform it into an infrared spectrum. Unfortunately, the infrared situation is complicated by the fact that the molecules don't re-emit the same frequencies they absorb, the way nuclear magnetic moments do. (The molecules redistribute the infrared energy through themselves as random thermal motion.) So we don't have the simultaneous emission from the molecule of multiple frequencies that interfere with each other to give an interference pattern that can be stored and transformed.

This doesn't prevent us from getting the IR spectrum, though. We generate the IR radiation and create an interference pattern in it *mechanically* within the instrument—then the molecules in the sample can absorb their frequencies out of the interference pattern. So when we computer-transform the final pattern, the Fourier transform will show those frequencies to be missing. That is, FT-NMR uses an interference pattern that's *internal* to the molecule, while FT-IR uses an interference pattern that's *external* to the molecule.

All right, how do we get an interference pattern in a beam of infrared radiation and have it change uniformly with time? Figure 15 gives a block diagram of the instrument, called an ***interferometer***, that does this. The *globar* source **A** is just a white-hot ceramic rod that glows, radiating infrared as well as visible radiation.

Figure 15

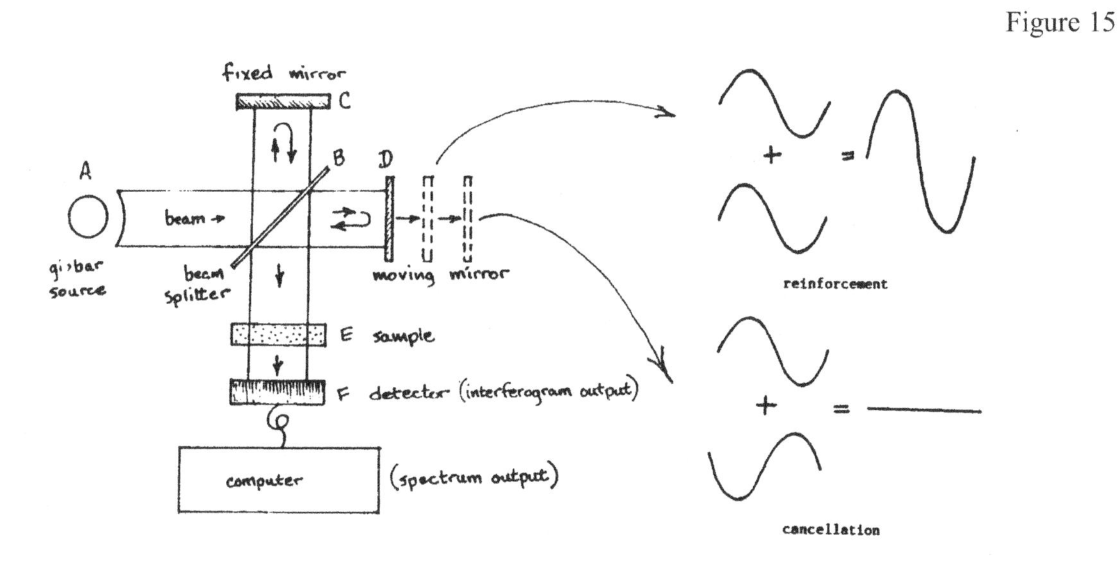

A beam of IR radiation from the source moves toward a ***beam splitter*** **B**, which is a half-silvered mirror (like silvered sunglasses) that allows half the radiation to pass straight through and reflects the other half off at a 90° angle. The reflected half strikes a ***fixed mirror*** **C** and bounces straight back, while the unreflected half strikes a ***moving mirror*** **D** that is slowly moving away from the beam splitter, and is also reflected back toward the beam splitter. The two half-beams recombine at the beam splitter and are reflected down toward the sample (half goes back toward the source, but we can ignore that).

What has happened to the recombined beam at this point?

At the moment the two mirrors are exactly the same distance from the beam splitter, so that the two halves of the IR radiation beam are traveling through exactly the same path length (the ***Zero Path Difference*** or ***ZPD*** condition), all wavelengths combine equally and the two halves of the beam "interfere" with each other by reinforcing each other, no matter what wavelength (or frequency) we look at. Figure 16 below shows this condition, but also shows a different situation in which the half-beam from the moving mirror is delayed by half a wavelength's travel *for a particular wavelength*, so that it combines with the half-beam

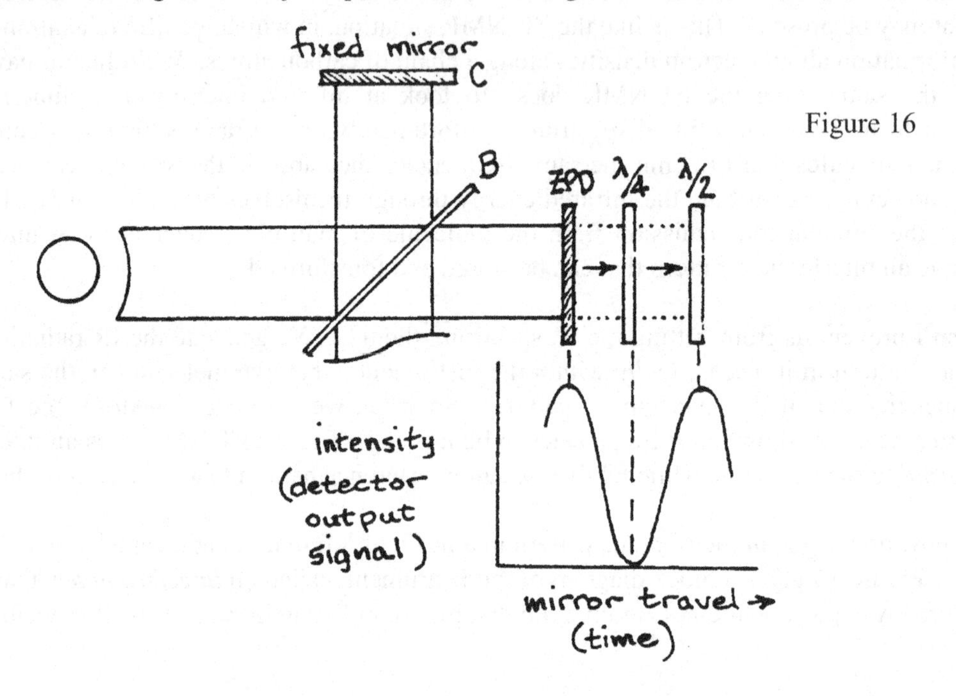

Figure 16

from the fixed mirror by genuinely interfering with it—canceling it, in fact. This is what happens when the moving mirror has stretched its path length out to an extra quarter of a wavelength (λ/4) for its half of the beam. That means that by the time its wave has gone out an extra λ/4 and back an extra λ/4, the total extra distance is λ/2—and half a wavelength is just enough to create the cancellation shown.

Of course, once you get away from the ZPD equal-path condition, a given mirror position will only cancel out one specific wavelength (or frequency); other wavelengths would have different degrees of interference. But if the mirror keeps moving, eventually it will reach the cancellation (λ/4) condition for every wavelength (or frequency) we're interested in. So the degree of interference for any particular wavelength will change systematically as the mirror travels. Figure 16 shows how combined-beam intensity depends on mirror travel for one particular wavelength; because the mirror travels at a uniform speed, this is equivalent to a graph of intensity versus time as the mirror scans. Aha! This sinusoidal graph of intensity vs. time is just like what we saw in Figure 10 for one given frequency (or wavelength). If another frequency is also present, we'll get the same kind of interference pattern we saw in Figure 10 for those two frequencies—and the globar source is giving us a whole range of frequencies. So we have indeed generated an interference pattern that's external to the molecule.

What will this mean to the intensity of the beam, with its thousands of interferences, as the moving mirror scans along? Because there *are* thousands of wavelengths, at any given mirror position about half will be reinforcing (high intensity) and half will be cancelling (low intensity). So the overall intensity will be just about 50%, regardless of the mirror's position – it's just that different wavelengths will be reinforcing and cancelling at each different mirror position. We'll get a graph of intensity vs. mirror travel (that is to say, intensity vs. time) like Figure 17 below – about 50% brightness everywhere except at ZPD where everything is in phase and at max brightness. This curve is called an ***interferogram***; it's analogous to the NMR's FID.

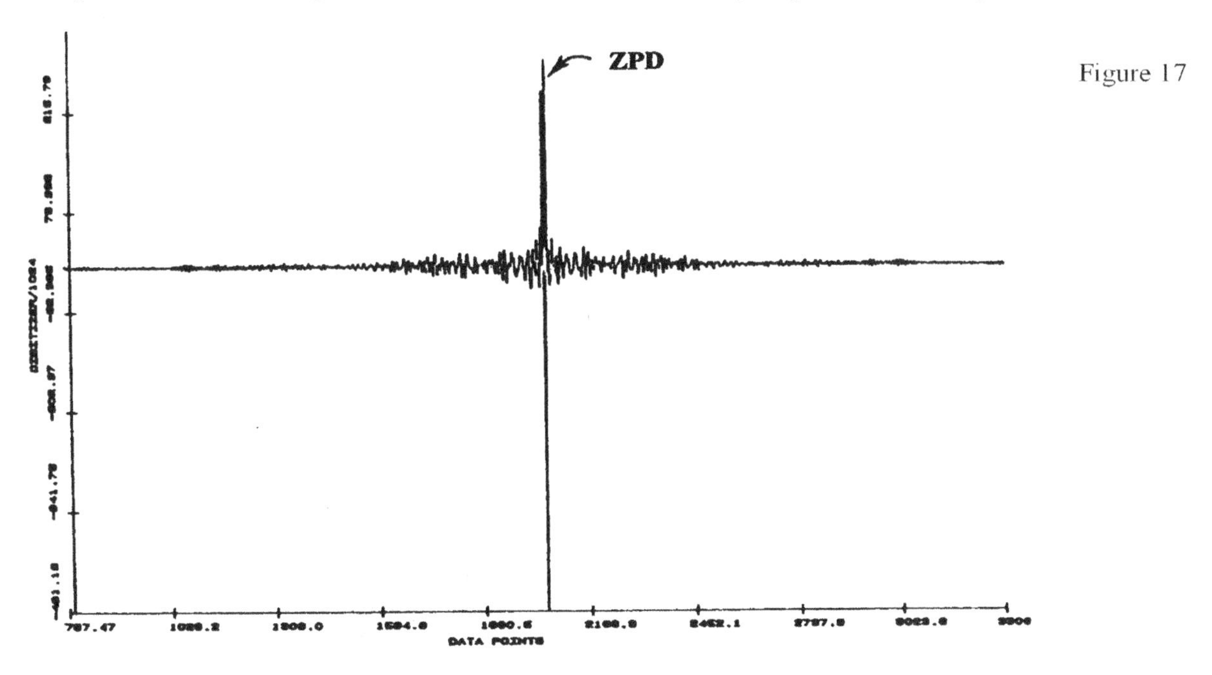

Figure 17

Why isn't the interferogram flat at 50% away from ZPD? Well, even with no sample in the instrument, there's still air inside it, and the air has water vapor and CO_2 that can absorb certain infrared wavelengths. That means those wavelengths won't be there at the detector even when the moving mirror says they should. Absorption by the air (or, of course, by a sample) puts hash on the 50% line. We'd like to avoid the air absorbances when we look at a sample, and we'll have to think about how to get rid of them in a minute.

At this point, then, we have an interference pattern in our beam of infrared radiation. We can now send the radiation on to the sample (**E** in Figure 15), which will absorb certain frequencies (those equal to its natural vibrational frequencies), thus modifying the overall interference pattern. The beam of IR radiation, with some frequencies absorbed out, goes on to the detector **F**, which produces an electrical output signal proportional to the total intensity of all the IR radiation striking it. A plot of this output signal as a function of time is the interferogram for our sample. As for the NMR instrument, we have a captive computer that does the work of producing the Fourier transform of the interferogram: total intensity of all frequencies as a function of time (mirror travel) is transformed into intensity as a function of reciprocal time (IR frequency), which is the IR spectrum we seek.

Well, almost.

When you run an infrared spectrum on the FT-IR, it will first record the interferogram for a series of 20 mirror scans, added together to reduce optical and electronic noise. Next, it will look at its navel momentarily while it does the Fourier transform of the cumulative interferogram. But what it shows you won't be the interferogram or even the immediate FT result. That would be a sort of bell-shaped curve like Figure 18, but with some hash (the atmospheric absorption).

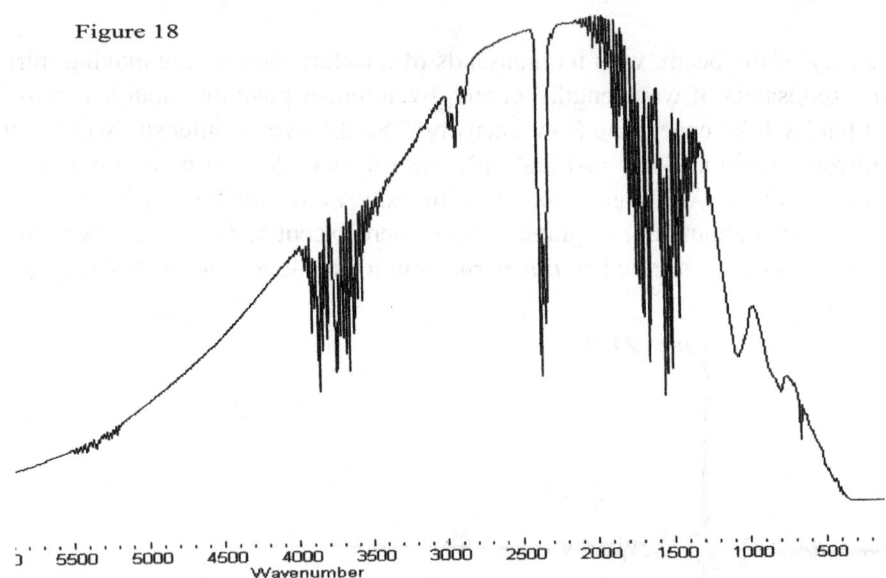

Figure 18

The bell-shaped curve is just the black-body emission curve for the original globar source, and what we have here is the ***background spectrum***. We'll want to compare it to the sample's spectrum. So we save the background spectrum and do the whole scan again with a sample in place. Figure 19 shows the result, but with both the background and sample spectra shown superimposed.

You can see that there are some wavelengths (or wavenumbers) where the sample is indeed absorbing, but we'd like to clean the spectrum up to do two things: (1) get rid of the atmospheric absorption peaks, and (2) jack the baseline up to a sort of flat line at the top of the paper – a flat baseline.

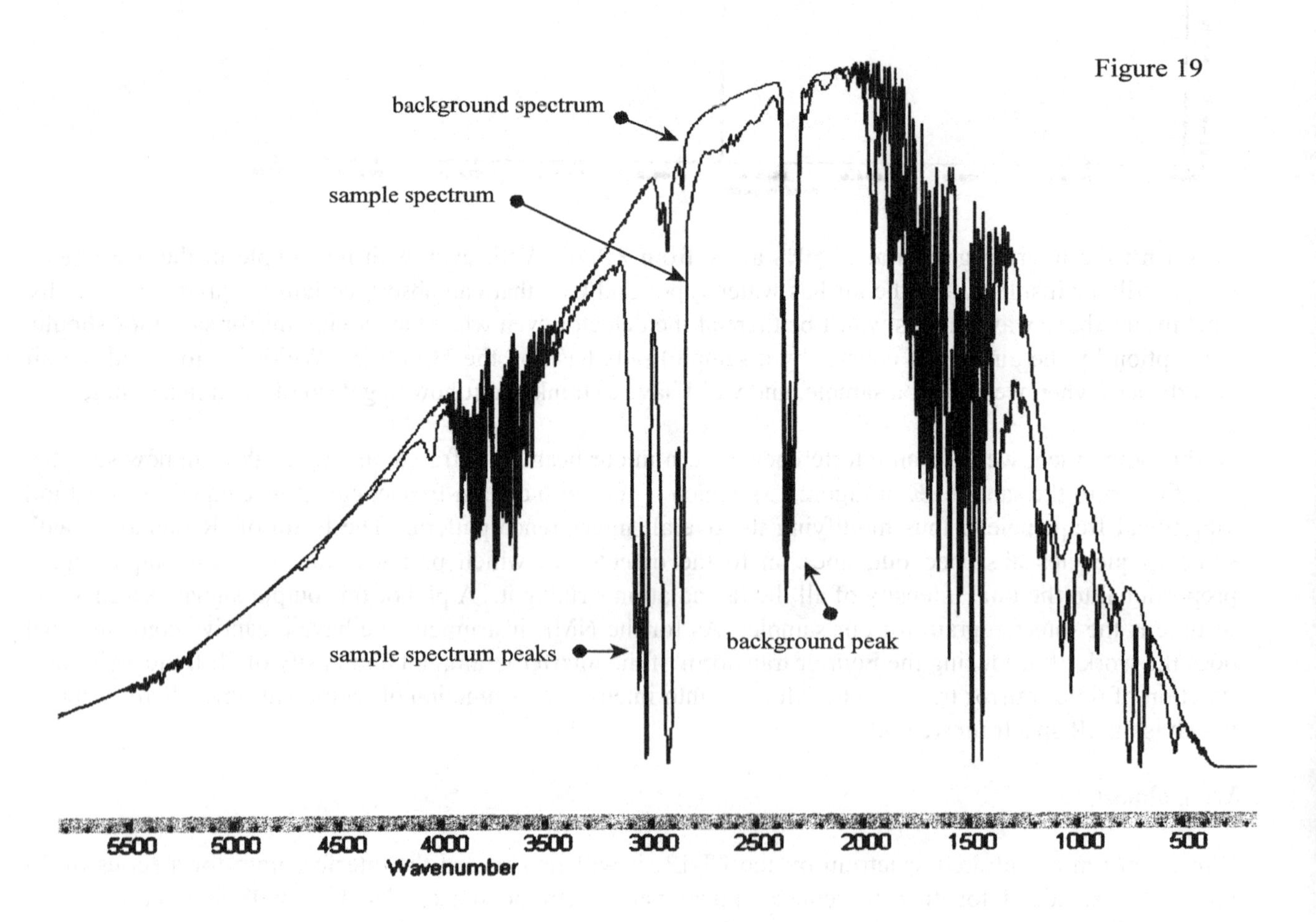

Figure 19

The instrument will do this for us automatically. What it does is to take the ***ratio*** between the sample spectrum and the background spectrum: sample/background. Out at the edges where the curve is very low the ratio will be, maybe, 1.8%/1.8% or 100%; in the middle the ratio will be, maybe, 44.2%/45.6% or 96.9%; and at a sample peak the ratio will be perhaps 2.2%/46.9% or 5.5% – so there's a baseline up near 100%, but peaks are still peaks. The result, for polystyrene (the sample in Figure 19) is shown in Figure 20.

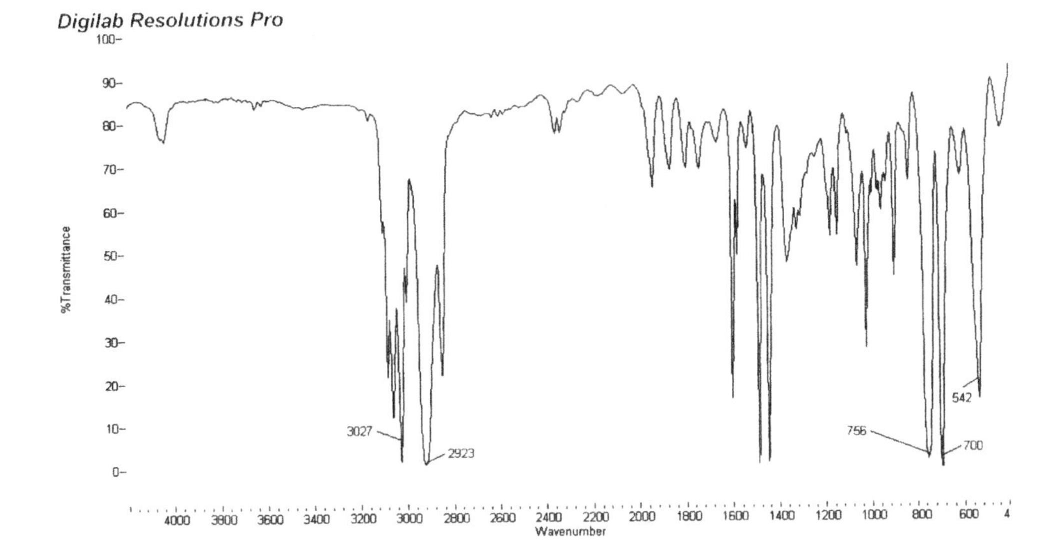

Figure 20

Note that you can have the computer index the wavenumber of peaks you pick out.

Briefly, then: A hot source of infrared radiation (a globar at ***A*** in Figure 15) shines a beam of IR frequencies onto a half-silvered mirror or beam splitter (at ***B***), which sends half the beam to the fixed mirror ***C*** and half to the moving mirror ***D***. These two halves of the original beam recombine by being reflected back to the half-silvered mirror again, go on (with their interference patterns already present) to the sample ***E***, which absorbs some energy out of the overall interference pattern and allows the rest to go on to the detector ***F.*** The detector gives us an interferogram (intensity as a function of *time*), which the computer turns into a spectrum (intensity as a function of *reciprocal time* or frequency) by taking the Fourier transform of the interferogram. Then the computer shows us the *ratio* of the sample spectrum to the background spectrum, which cleans it up and gives us a flat baseline.

The FT-IR Technique Quiz

Once you've read through the material above and your notes from the IR lab lectures, and asked questions till you understand the technique, sit down with your notebook and the short-short list of candidate compounds that

1) have about the right boiling point to match your unknown, and
2) have a consistent ^{13}C NMR spectrum (right number of peaks *and* right chemical shifts).

(There might be three or four of these, but you could equally well be down to one candidate at this point.) For each candidate compound, draw out its molecular structure (an expanded version of the stick formulas in the table, showing each C atom as the letter C and showing all hydrogens) so as to show all the bonds present. Then, using the discussion above on how to use the correlation chart, work out where the important absorption peaks should be for that structure (all the peaks above 1500 cm-1 and the important ones below that frequency) and what kind of natural vibrational mode is responsible (from the bottom of the chart).

Make out another little table, working down from high frequency to low, listing these peaks; for our earlier example of 3-pentanone, we would have

	Absorption pks	Vibrational mode
O	3000-2830 cm-1 (3 pks)	C-H stretch
‖	1730-1650	C=O stretch
CH_3-CH_2-C-CH_2-CH_3	1480-1430	C-H bend

Set up a separate table like this for *each candidate*, and if you have more than one candidate remaining, add a short discussion of exactly *which* peaks should distinguish one of these from another. If you want to, you can look up the spectra of your candidates in the *Aldrich FT-IR Library* in the classroom for help in this distinction—but **get the data for the little table from the wavenumber correlation chart, Table V, and from the table of key wavenumber ranges**. Your candidates *must* be based on complete ^{13}C NMR interpretation, of course.

When you've written this out in your notebook and have a reasonable idea about how the technique works, bring your notebook to a faculty member and ask to take the IR quiz. That quiz will have the same format as the one for ^{13}C NMR, in that there are two questions about the technique that you must answer out of your head, with no notes, and a third question about your candidates that you can answer with the help of your notebook. The questions will be:

1) What happens when IR radiation hits a molecule?

This involves the dipole and electric field discussion, with the idea of frequency matching. Be able to give a short description (about two sentences) without notes.

2) How does the FT-IR instrument tell us what frequencies of IR radiation are being absorbed?

Study the section on FT-IR Instrumentation, particularly the block diagram, and be able to work your way through the spectrophotometer's sequence of events, starting with the source and winding up with the transformed spectrum. No notes here either.

3) Exactly where are your candidate molecules going to absorb?

If you've properly written up the little tables in your notebook, you can simply show that discussion to the instructor—but the tables should be complete in the format shown.

At this point you should have your notebook up to date through the ^{13}C NMR, and the instructor will check that too.

Running the IR Spectrum

Once you've passed the oral quiz and have the instructor's initials in your notebook, take the two bottles with your low-boiling and high-boiling compounds to the IR room (212). Get a lab assistant to help you set up a sample on NaCl plates ***or*** on KBr plates if you think you might have a Br-substituted compound. Then you can run the Digilab instrument yourself using its Windows software:

1) **Double-click on DIGILAB RESOLUTIONS PRO icon, wait for blank screen in top half**
2) **Make sure nothing is in sample compartment**
3) **Click on COLLECT in top bar, then on RAPID-SCAN in drop-down list**
4) **Check that RESOLUTION = 4 cm^{-1}, and SAVE RANGE = Custom 4200-400 cm^{-1}**
5) **Click BACKGROUND at bottom, wait for 20 scans**
6) **Save to BACKGROUND file, replacing old background**
7) **Open sample compartment, insert KBr disc holder with sample prepared by instrument tech**

8) Click on COLLECT in top bar, then on RAPID-SCAN, then on SCAN, wait for 20 scans
9) Both background and spectrum will be on screen; click on bottom miniature spectrum in lower list, then on top miniature spectrum to display spectrum alone in top screen
10) Click on BACKGROUND(2) in list to highlight word; type in your name and "HB" or "LB" to identify spectrum
11) Drag horizontal bar between spectrum and lower list down to hide list
12) Move mouse to spectrum, where it should have an attached magnifying glass; move to bottom tip of first important peak, right-click, highlight NEW, click PEAK to attach a wavenumber label. If necessary, mouse-drag the label away from other peaks. Repeat for other important peaks, perhaps 6 to 8 in all, paying particular attention to peaks your NMR has suggested should be there. Finally, left-click another peak to remove all highlights.
13) Type Ctrl-P to print; click OK in printer panel for printout
14) Click on OPERATIONS in top bar, then highlight SPECTRALID in drop-down list and click it. Wait for the black logo square to disappear from the screen. Click the SEARCH button in the top row, which has a magnifying glass over a red spectrum and a book.
15) Click the PRINT button in the top row at the left, which shows a printer. When the printer panel opens, click PRINT at the bottom left.
16) Close SpectralD by clicking the exit × button at the right corner of the top Windows bar
17) Repeat steps 7-16 with your other unknown, with the instrument tech preparing the sample

Notebook Writeup

This is a new experiment, of course; use the usual headings:

- **Purpose (step 14, page 30)**
- **Procedure (step 15)**
- **Apparatus (step 16 — include instrument: Digilab Excaalibur FT-IR spectrophotometer)**
- **Data (step 17 — spectrum and search on LH pages, table and interpretation RH page)**
- **Summary and Conclusions (step 20)**

Mount (1) your low-boiling spectrum and search together on a left-hand page (probably p. 36) right side up, taped at the left and folded once to fit inside, then (2) your high-boiling spectrum and search on the next left-hand page (probably p. 38). On the opposing right-hand pages, interpret the spectra, paying particular attention to peaks your previous NMR suggested should be there. Use the new right-hand page (***Data***) to interpret the peaks you've indexed. Identify the probable general vibrational mode ("C=O stretch", for instance) and also the specific fit for a molecule on your candidate list, comparing it with your earlier little tables. Spend a short paragraph discussing which of your candidates is the best fit and why; remember that missing peaks are as important as those that are present! It is not adequate to simply say that your experimental spectrum is identical to the library spectrum for 3-pentanone, or whatever! *You must interpret the individual peaks to receive full credit.*

Mount the printed set of library spectra and list of search matches in your notebook on a new left-hand page. Although you can't index the peaks as you did for your experimental spectrum, you can read the wave-number values roughly with a ruler from the bottom scale. Use the presence *or absence* of key peaks to reject all but one, or at most two, of your candidates. Under the ***Data*** heading, provide a discussion in your notebook of what peaks lead you to your conclusion (see below). If all your candidates seem to be ruled out, consult a faculty member; you might need to request some more comparisons from the spectral library. Finally, under ***Summary***, give a brief discussion of where your search stands now, with particular emphasis on the new information IR has given you (which possibilities were discarded, etc.).

^{1}H (PROTON) NUCLEAR MAGNETIC RESONANCE (NMR) (after quiz, 30 min.; notebook p. 45)

Thus far we have discussed only ^{13}C NMR and have treated the substituent hydrogen atoms as nuisances. Since the hydrogen nuclei also have nuclear spins and behave as if there were little magnets embedded in their nuclei, you shouldn't be surprised to learn that we can also do ^{1}H NMR spectroscopy. Proton NMR differs from ^{13}C in three important respects:

(1) ^{1}H nuclei are much more common than ^{13}C nuclei (^{1}H has an isotopic abundance of 99.9844%);

(2) ^{1}H nuclei are about 60 times more sensitive to radiated energy than are ^{13}C nuclei; and

(3) the range of ^{1}H chemical shifts, 10 ppm, is about twenty times smaller than that for ^{13}C nuclei.

Consequently, ^{1}H NMR is easier to do (abundance and sensitivity) and in fact was developed first—but the information is crowded into a smaller spectral range (narrow chemical shift range). Further, hydrogen-hydrogen ***spin-spin coupling*** always increases the complexity of ^{1}H spectra, because essentially all hydrogen nuclei are protons and have the same nuclear spin. Let's look at the familiar chemical-shift part first, and come back to the new topic of spin-spin coupling.

Table VI on the opposite page presents typical proton NMR chemical shifts that result from various molecular functional groups. There is a rough correlation with ^{13}C shifts because the same shielding mechanisms operate in both kinds of NMR. One fairly modest difference between proton spectra and the earlier ^{13}C spectra is that, because the total chemical shift range is so much narrower for proton spectra (10 *ppm* vs. 200 *ppm*), the individual peaks will appear broader when the spectrum is spread out to fill the chart paper. But you can predict chemical-shift ranges for proton NMR spectra just as you did for ^{13}C spectra.

Well, the general chemical-shift behavior is the same for proton NMR spectra as for ^{13}C spectra—but what about the spin-spin coupling business? When neighboring nuclei (connected by a bond between the atoms) both have a spin (nuclear magnetic moment), the moment from one nucleus becomes part of the total magnetic field experienced by the other nucleus. This has the effect, in the spectrum, that a given peak will be split into several evenly-spaced peaks by the presence of its neighbor spins— the spectrum looks a lot more complicated, but it contains a lot more information about the structure of the molecule. This interaction, which arises from coupling between the nuclear magnets (***"spins"***), is called ***spin-spin coupling***.

Unlike the electrons that circulate throughout a volume of space, the nuclei are held at fixed positions within the molecule by chemical bonds. Thus the magnetic field set up by a neighboring *nuclear* magnet has a definite value and, since the neighbor magnet is also aligned either parallel or antiparallel to the NMR's magnetic field, the nucleus we're interested in sees an added field either assisting or opposing the external field. We see the two possible results that can occur on the page after Table VI:

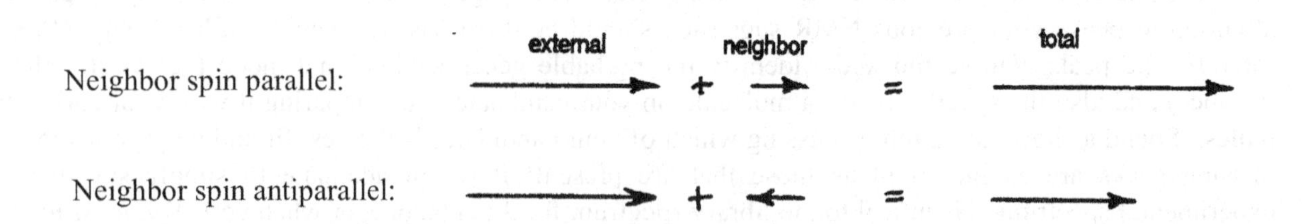

Since the *total* magnetic field seen by the nucleus we're studying must be just large enough to satisfy the resonance condition, we see that there will be two values of the external field at which resonance will occur: a smaller value for the neighbor-spin-parallel case and a larger one for the neighbor-spin-antiparallel case. The original single resonance NMR line of the nucleus isolated from any neighbor influences is seen to split into two resonance NMR lines (each half the original height) when a single neighbor spin interacts.

Table VI — Typical ¹H NMR Chemical Shifts vs. TMS

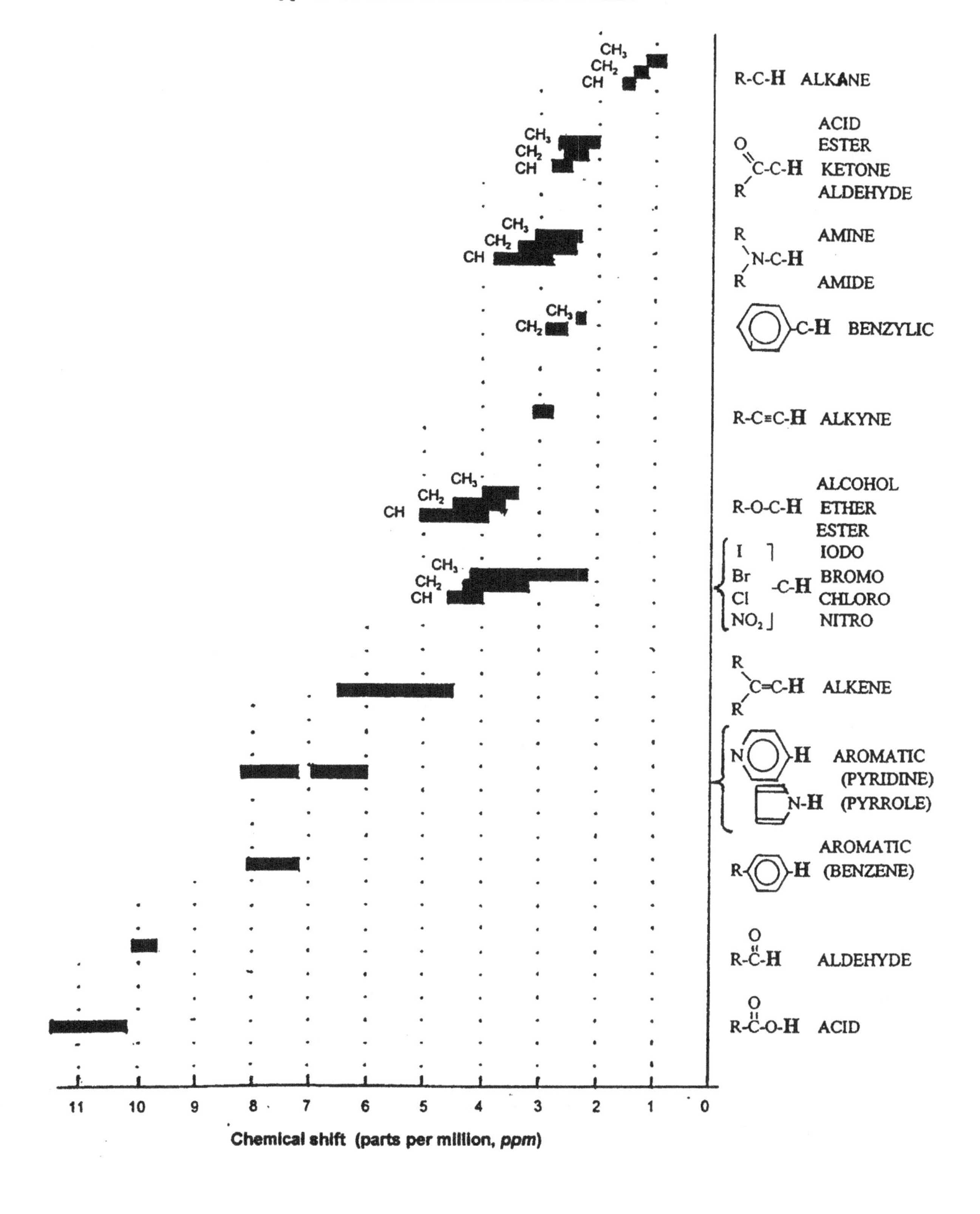

What we've said so far isn't specific for proton nuclei. This interaction is going to occur any time two nuclei with spins are bonded to each other. Why didn't we see it occur in the ^{13}C NMR spectra, where a carbon nucleus with spin was bonded to several hydrogen nuclei with spin? Well, it would have been there, but we electronically scrambled the orientations of all the hydrogen spins so that their magnetic effect just averaged out to zero, which simplified the spectra you saw. This process is called ***decoupling***; your ^{13}C NMR spectra were proton-decoupled to eliminate this effect. But if they hadn't been decoupled, the peak due to a carbon with one hydrogen on it (a -CH= group) would have been split into two peaks (a doublet).

What if there are two nuclear-magnet (spin) neighbors, as in a $-CH_2-$ group? Extending the argument above, we see that three cases occur: (1) both neighbor magnets are parallel to the external field, (2) one neighbor spin is parallel while the other is antiparallel, thus cancelling each other's effect, and (3) both neighbor magnets are antiparallel to the external magnetic field :

	external		neighbors		total
Both parallel:	⟶	+	→ →	=	⟶
Opposite spins:	⟶	+	{ → ← ; ← → }	=	⟶
Both antiparallel:	⟶	+	← ←	=	⟶

In this case there are three transitions instead of the original (unsplit) single line. However, since there are two different combinations that have one spin parallel and the other antiparallel (with no net effect on the resonance field), the center transition is twice as intense as the outer ones at higher and lower field—1:2:1 intensity ratios. Similar arguments should lead you to conclude that three neighbors (as in $-CH_3$) will split the original ^{13}C line into four transitions of relative intensities 1:3:3:1. This is a general result:

> ***N equivalent hydrogen neighbors will produce N+1 transitions*** and their intensities will follow the pattern of the binomial expansion coefficients (***Pascal's triangle***).

Of course, carbon atoms can only have 0, 1, 2, or 3 hydrogen neighbors directly attached—so we have considered all the possibilities.

Well, that's what would have happened to the ^{13}C peaks if they hadn't been decoupled for the sake of simplicity. What about the spin-spin coupling in proton NMR spectra, which is the technique we're taking up here? Hydrogen atoms can only form one bond, and that will usually be to a carbon atom in our organic compounds. We don't have to worry about the proton peaks being split by the ^{13}C spins, because only 1% of the carbon atoms have mass 13 and a spin. That's so few that we can afford to ignore the chance of a proton peak being split by a neighbor ^{13}C. However, the effect of neighbor-proton spins extends as far as the hydrogens on the *next carbon atom over* in the structure. So if we had the structure group =CH-C**H**= and were studying the right-hand hydrogen (proton), its absorption peak would be split into *two* peaks by the left-hand proton's spin. If we had the structure $-CH_2-C$**H**= and were studying the right-hand hydrogen, its peak would be split into *three* peaks by the two equivalent left-hand protons. And conversely, if we were studying the two left-hand hydrogens, their peak would be split into *two* by the right-hand proton.

The only hitch to remember is that symmetrically equivalent protons don't split each other. For example, in a $-CH_3$ group the three equivalent protons don't split each other; and in ***ethane***, CH_3CH_3, the two CH_3 groups are equivalent and they don't split each other either. The proton NMR spectrum of ethane is a single peak.

Proton NMR spectra have one more advantage over ^{13}C spectra, in that the intensities of the peaks are more reliably proportional to the number of nuclei doing the absorbing. You may have noticed in your ^{13}C spectra that peaks representing a single carbon were about the same height, and those representing two carbons were about twice as high—but you may not have noticed it, because it doesn't always work out that way. Generally, only carbons that are chemically very similar will have similar peak intensities. The unreliability of peak intensities for ^{13}C NMR peaks is the reason we didn't mention intensities in the earlier discussion. But for proton NMR the situation is simpler: the intensities of proton NMR peaks are proportional to the number of protons doing the absorbing, within about 5% error. Just as for gas chromatography, the intensity of a peak is measured by the area under it, not by its height; and the *absolute* intensities (areas) don't matter, only the *ratio* of intensities for two or more peaks that occur on the same spectrum (chromatogram).

So if your proton NMR spectrum has two or more peaks, you should expect to measure the areas under the peaks and compare the area *ratios* to see if they match the stoichiometry ratios in your candidate compound. This is done by electronically integrating the peak height to give the area, and the NMR instrument allows you to request ***integration*** for two or more peaks.

So three kinds of information about the molecule's structure are found from its proton NMR:

(1) What kind of electronic environment is each absorbing group of hydrogens exposed to? (Found from the chemical shifts of the proton NMR peaks and careful reference to Table VI.)

(2) How many hydrogen atoms are absorbing at each chemical shift? (Found by integrating the areas under the proton NMR peaks and comparing the areas if there are two or more peaks.)

(3) How many neighbor hydrogen atoms are one carbon atom away from the parent atom of the absorbing hydrogen atoms? (Found by counting the number of lines into which the absorption is split by spin-spin coupling, and applying the N+1 rule in reverse.)

Now let's look at a few proton NMR spectra and try to see how each of these three pieces of information fits into the molecular structure analysis. Figures 20 and 21, on the next page, show the proton NMR's of ***nitromethane***, CH_3NO_2, and ***acetone*** $(CH_3)_2C{=}O$. Remember that the NMR absorption shown is of the *hydrogen atoms* in these molecules. Since each molecule has only one kind of hydrogen atom, the spectra show single lines. The absorptions occur at different chemical shifts because of the different electronic influences of the nitro- and carbonyl- substituents. Note that the peak seems to be the same height in the two spectra even though acetone has twice as many CH_3 groups; this is a result of autoscaling by the NMR instrument. You can only compare peak areas within a single spectrum, and then only if the entire spectrum has been recorded at the same instrument settings.

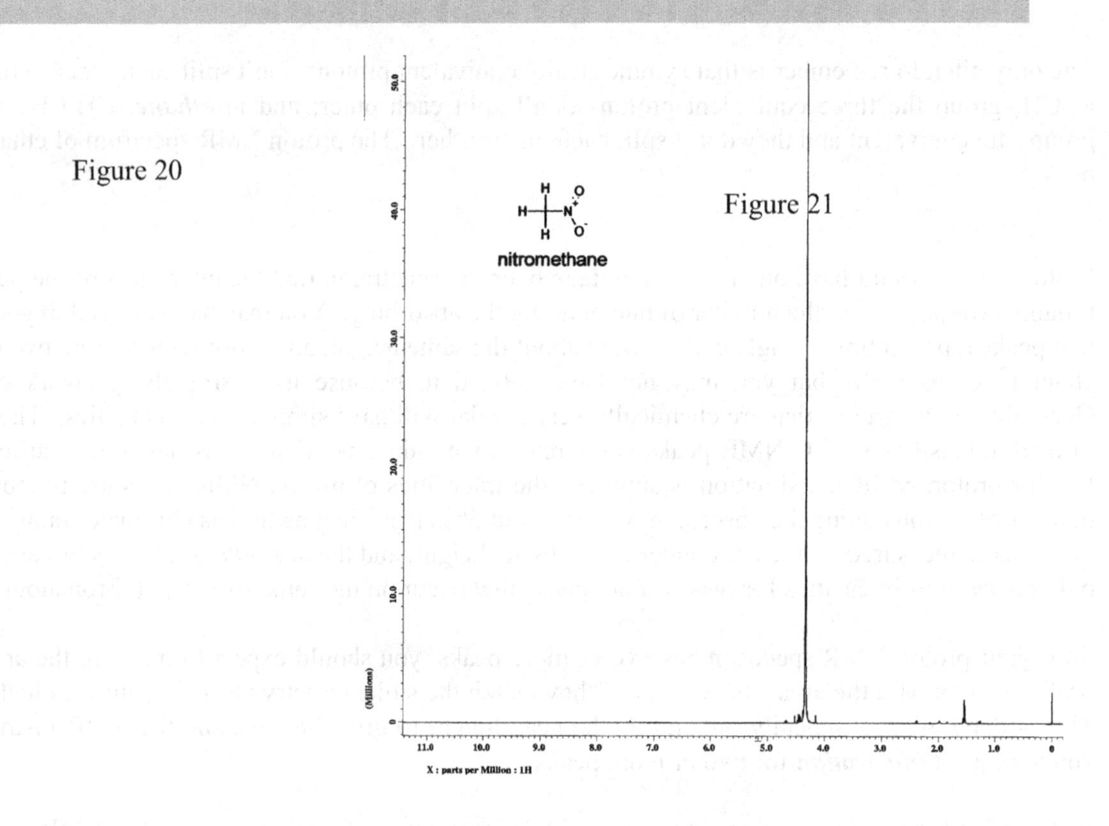

Figure 20

Figure 21

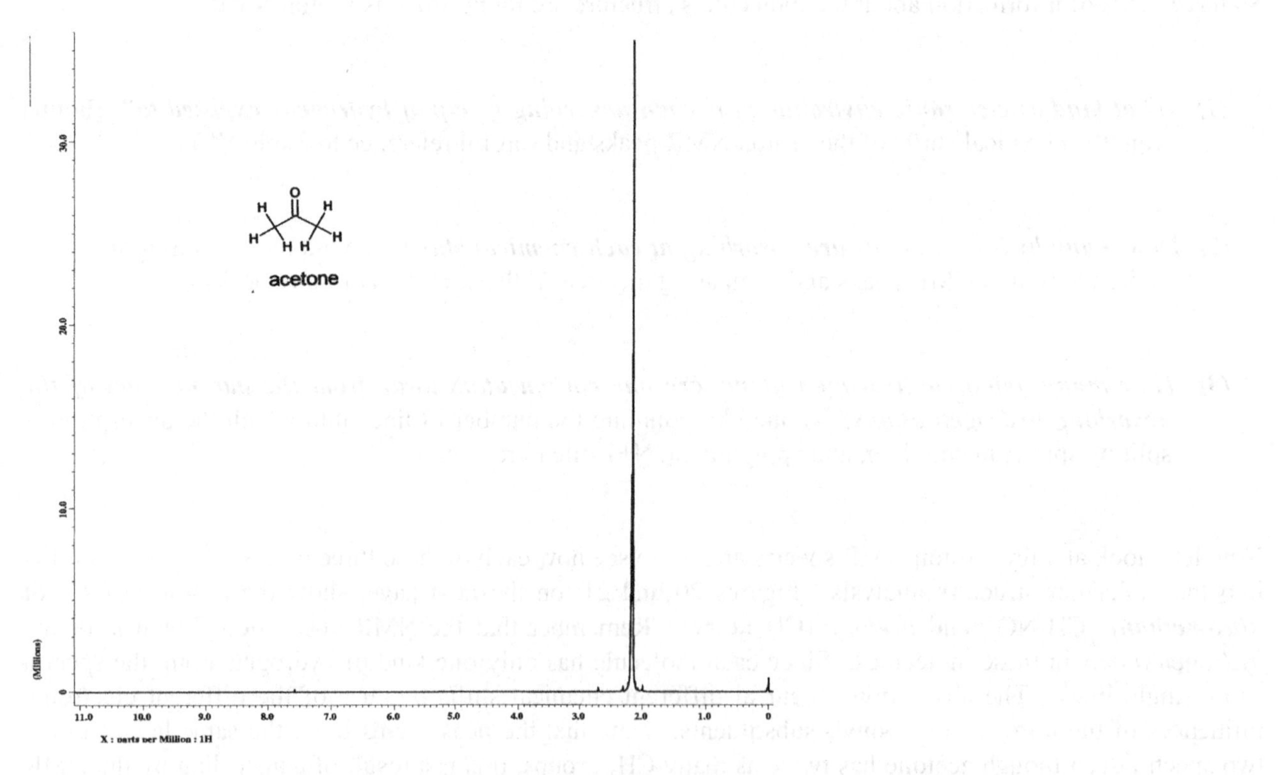

Figures 22 and 23 show the usefulness of integrated peak areas in assigning structures when several peaks are present: ***Toluene*** (Figure 22) has three methyl hydrogens and five benzene-type (aromatic) hydrogens. (On the spectra these areas are indicated by A3 and A5.) But **ortho-*xylene*** (Figure 23) has 6 methyl hydrogens and four aromatic hydrogens, and the line intensities reflect the differences. We can distinguish the two spectra—and molecular structures—even though the chemical shifts are very similar. Line intensities, especially when integrated, are a good guide to relative numbers of hydrogen atoms in proton

NMR, in contrast to line intensities in ^{13}C NMR (which don't say much except for small peaks for C's with no H bonded to them).

Figure 22

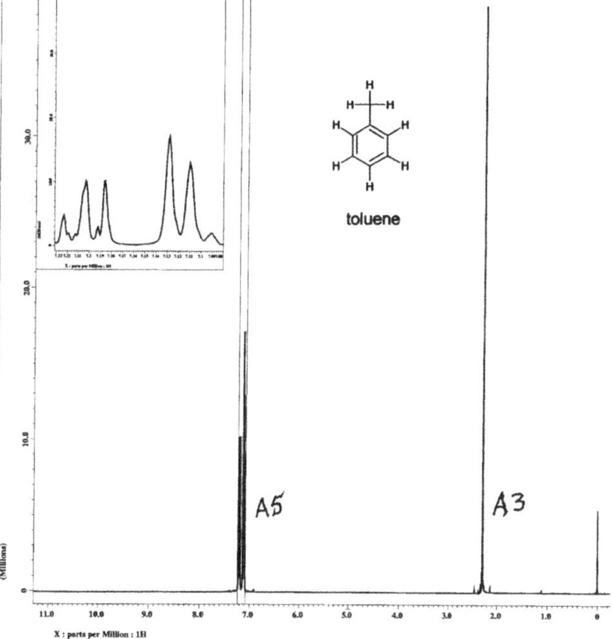

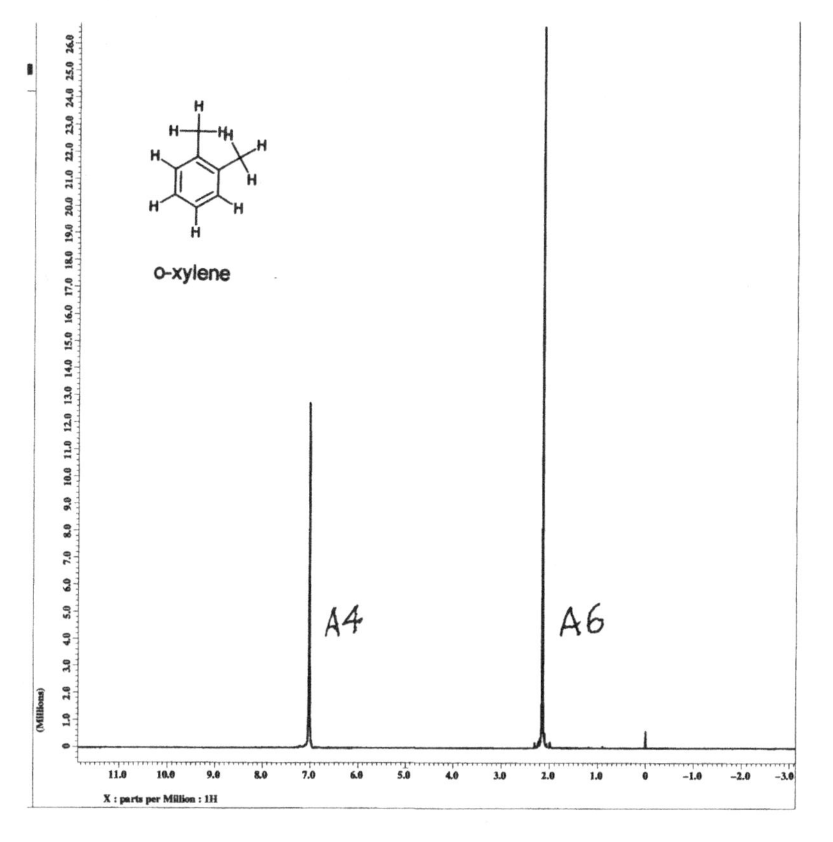

Figure 23

Figures 24 and 25 on the next page show how proton NMR lets you use chemical shift information to distinguish isomers of a complex ester. When the isolated *methyl* group is next to the -O- atom (Figure 25) it has a chemical shift of 3.6, greater than any of the ethyl (CH_3CH_2-) absorptions. However, when the *ethyl*

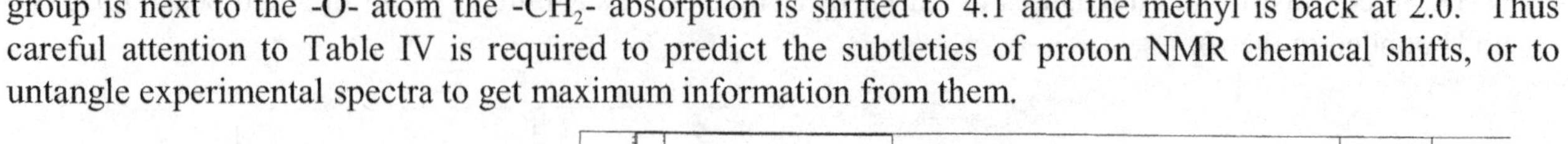

group is next to the -O- atom the $-CH_2-$ absorption is shifted to 4.1 and the methyl is back at 2.0. Thus careful attention to Table IV is required to predict the subtleties of proton NMR chemical shifts, or to untangle experimental spectra to get maximum information from them.

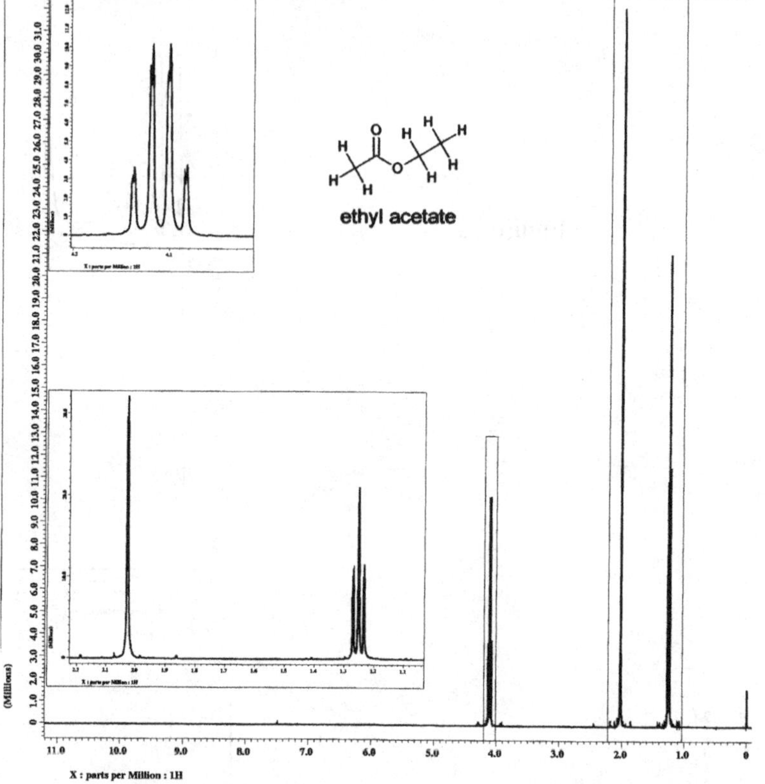

Figure 24

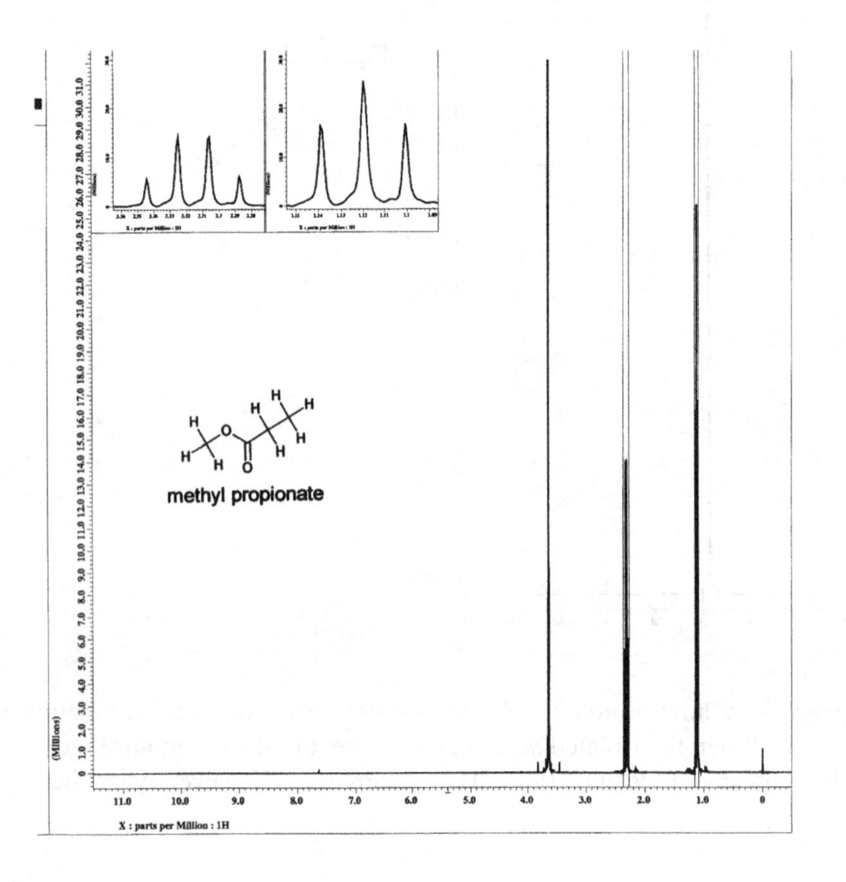

Figure 25

Figures 26 and 27 below show the complications that can arise when parts of a spectrum overlap each other. In Figure 26, the proton NMR of ***ethyltoluene*** shows near-overlap of the single CH_3 resonance of the methyl group with the $-CH_2-$ resonance of the ethyl group. (Note that the $-CH_2-$ peak is split into a 1:3:3:1 pattern, called a ***quartet***, by its own $-CH_3$ neighbor group. In a complementary fashion, the $-CH_3$ peak at 1.17 ppm is split into a 1:2:1 *triplet* by its neighbor $-CH_2-$ g roup.) The quartet and triplet "belong together" in that they're linked by the molecular structure, but the *singlet* from the methyl group is between them.

Figure 26

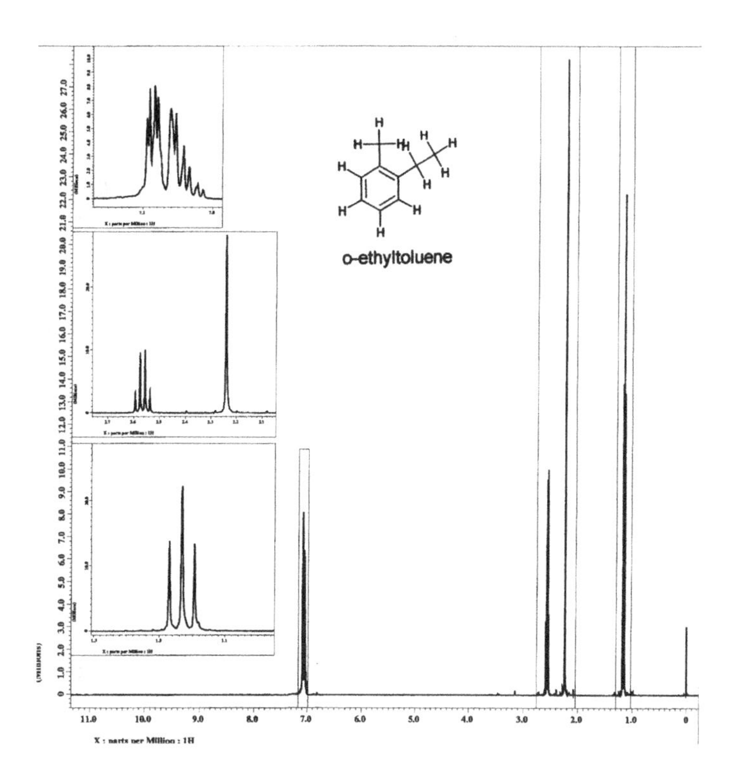

A more subtle complication is seen in Figure 27, the proton NMR of ***isopropyl bromide*** or *2-bromopropane.*

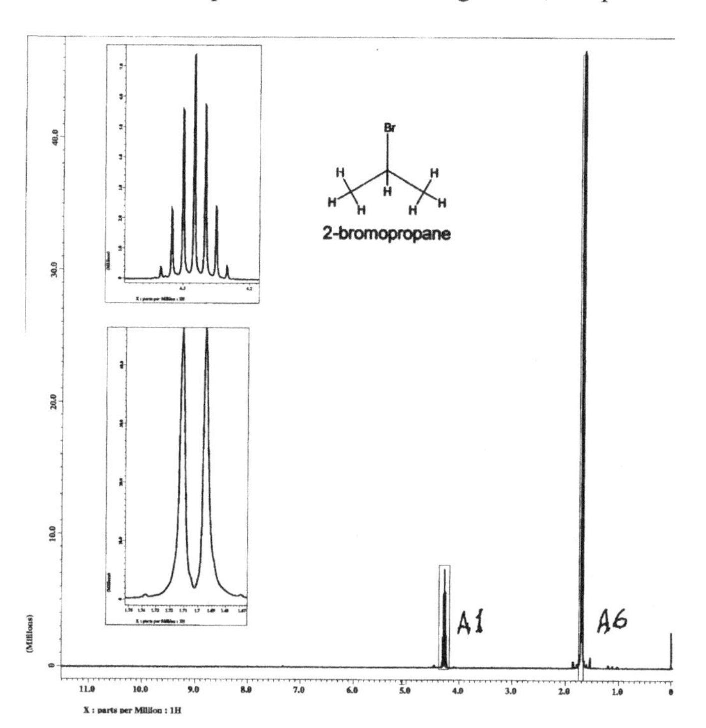

Figure 27

This spectrum shows only two sets of peaks: a *doublet* of large intensity and a *septet* of low intensity. The doublet is from the two identical -CH_3 groups, split by the single H on the neighbor CH group; the septet is from the lone CH, split by all six of the protons on the two neighbor CH_3 groups. These two spectra illustrate two useful general patterns:

1) an isolated ***ethyl*** group will show two peaks, a *quartet with area 2 and a triplet with area 3*
2) an isolated ***isopropyl*** group will show two peaks, a *small septet with area 1 and a big doublet with area 6.*

The ^{1}H NMR Quiz

Once you have the hang of the additional information that peak intensities and splittings (spin-spin coupling) can give you in a proton NMR, write up in your notebook the predicted proton NMR spectra for the compounds that remain candidates for each component of your unknown. Just as you lettered carbon atoms in terms of their symmetry type as *a,b,c...* before, sketch the structural formula for the molecule and add in the specific hydrogen atoms *a,b,c...* here. Set up a little table **for each candidate molecule** that looks like this:

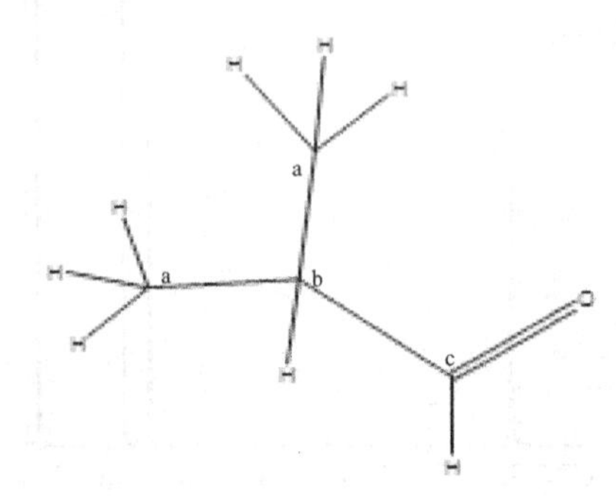

H type	chem. shift	rel. area	neighbor H	splitting
a	0.8-1.2 *ppm*	6	1	2
b	2.5-2.8	1	7	8
c	9.7-10.1	1	1	2

When you have this table set up for each candidate compound, find a faculty member and ask to take the quiz on proton NMR. The quiz will consist of the following three questions:

1) What are the three pieces of information that you can get from the proton NMR spectrum of a compound? The answer to this will involve the fundamental atomic behavior features (coupling of nuclear spins, etc.), the way the spectrum will be influenced (splitting of a single peak into a N+1 multiplet, etc.), and the molecular structure information (number of neighbor-carbon hydrogen atoms, etc.). You'll need to give these three components of an answer for each of the three pieces of information.

2) Predict the proton NMR spectrum of ethyl bromide, CH_3CH_2Br, including the specific features to be expected for each of the three pieces of information given in question 1. Work this out and be able to generate a little table analogous to the ones you've written up for your candidates—but you'll have to answer this question and question 1 without notes.

3) Predict the proton NMR spectra for each of your candidate compounds. For this one, you can use the tables you've written up in your notebook. Just show us the notebook.

When you've passed the quiz, we'll initial your notebook and you can sign up for instrument time with the instrument technician. The notebook writeup should be like that for the ^{13}C spectrum, with reasons for ruling out structures and specific discussion of the fit for the successful structure assignment. There is a compilation of proton NMR spectra by Aldrich, analogous to the *Aldrich FT-IR Library*, but it's only available in book format, not as a computer data base. You can use it, of course, but remember that it's *not* an adequate discussion to say that your spectrum looks just like Aldrich 813C or something.

FIGURING OUT STRUCTURES WITH BENZENE RINGS FROM ^{1}H NMR SPECTRA

The previous discussion and examples (and Table VI) have made it seem as though hydrogen atoms on a benzene ring will always have a chemical shift somewhere between 7 and 8 ppm, and that they won't be split by their neighbor H atoms on the ring. That's an oversimplification, and in fact you can get information about the geometry of substitution on the ring *and* information about the approximate electronegativity of the substituted groups from the details of the chemical shift and splitting of the basic "aromatic" peak in the 7-8 ppm vicinity.

The reason none of this showed up in Figures 22, 23, or 26 is that those spectra are golden oldies, run on an old low-frequency NMR spectrometer. It turns out that the higher RF frequency you use (see Figure 8 on page 84), the better separation of peaks you get in the spectra. A current state-of-the-art instrument like the one we use will show the small splittings clearly for the 7-8 ppm aromatic-H peak, and even show different chemical shifts for the symmetrically different H's.

As a first example, consider a monosubstituted benzene ring (the basic $\mathbf{C_6H_6}$ ring with one group replacing a H so that the formula is now $\mathbf{C_6H_5}$-X). The five H atoms on the ring should fall in three groups with different chemical shifts: 2 on the next C to X, 2 with one C in between, and 1 with two C in between. If X is a carbon atom or something else with a fairly low electronegativity that doesn't suck electrons out of the ring, these three chemical shifts will all be very nearly the same and the peaks will overlap even on a high-frequency spectrum. There will be very small splittings: the first group above have one neighbor H and will be a doublet, while the other two groups have two neighbors and will be triplets. However, the nearly identical chemical shifts for the three groups make it hard to pick out the splittings; only two basic peaks show up, at about 7.3 and 7.4 ppm. Figure 28 below shows the spectrum (only the part between 6 and 9 ppm, blown up) for **toluene** or methylbenzene, $\mathbf{C_6H_5}$-CH_3 and you can see these features.

Figure 28

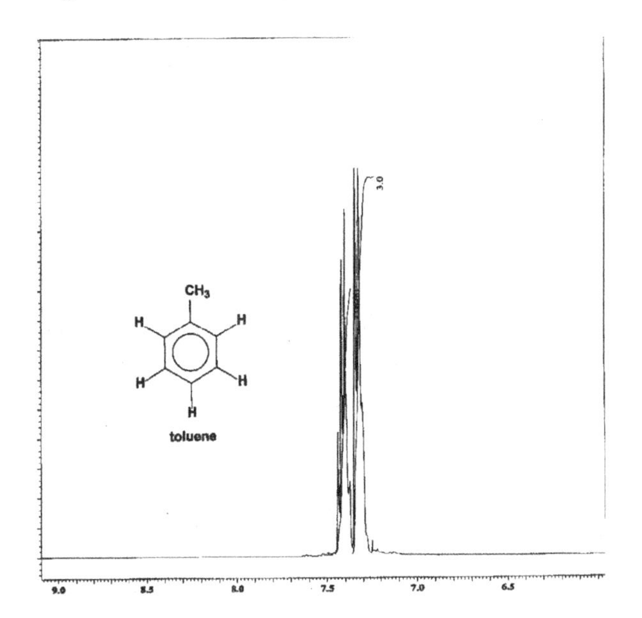

Figure 29

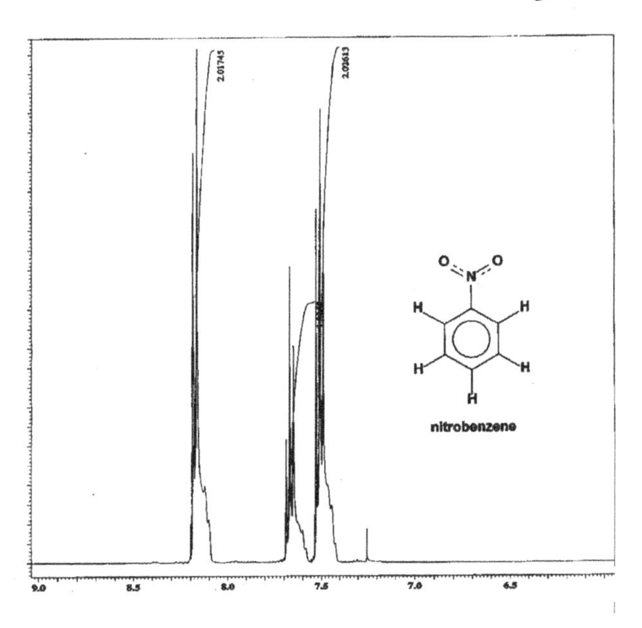

What changes if the substituting group has a high electronegativity and pulls electrons out of the ring strongly? Figure 29 shows the 6-9 ppm region of the spectrum of **nitrobenzene,** $\mathbf{C_6H_5}$-NO_2, where the -NO_2 group is quite electronegative. The primary difference from the toluene spectrum is that the strong electron withdrawal now gives the three groups of H atoms substantially different chemical shifts: instead of overlapping peaks about 7.4, there are now recognizable peaks at 7.50, 7.65, and 8.15 ppm. The doublet at 8.15 is due to the H's closest to the electron-withdrawing NO_2 group, and its chemical shift is affected most.

So you can distinguish an electron-withdrawing substituent group from an alkyl group easily by looking at the aromatic region of the ^{1}H NMR spectrum.

Suppose the benzene ring has *two* groups on it, so that the formula is Y-$\mathbf{C_6H_4}$-X. There are three structural possibilities, as the table on page 75 and the discussion just before it have indicated: *ortho-* or *o-* (groups next to each other on the ring), *meta-* or *m-* (groups one carbon apart on the ring), and *para-* or *p-* (groups opposite each other on the ring). Consider the *ortho-* structure first.

If the two groups are identical or at least very similar in electronegativity, symmetry indicates that there will be two groups of H's: next to the substituents, or one carbon away from the substituents. Figure 30 below shows the spectrum of ***o*-xylene**, CH_3-C_6H_4-CH_3, in the 6-9 ppm region. In this case it's hard to see that there are two peaks present, because the chemical shifts are so similar, but if the substituents were more electronegative as in ***o*-dichlorobenzene**, ClC_6H_4Cl, the two peaks would be at clearly different chemical shift values. When two peaks *are* visible they'll have the same area, because they each represent 2 H atoms.

Figure 30 Figure 31

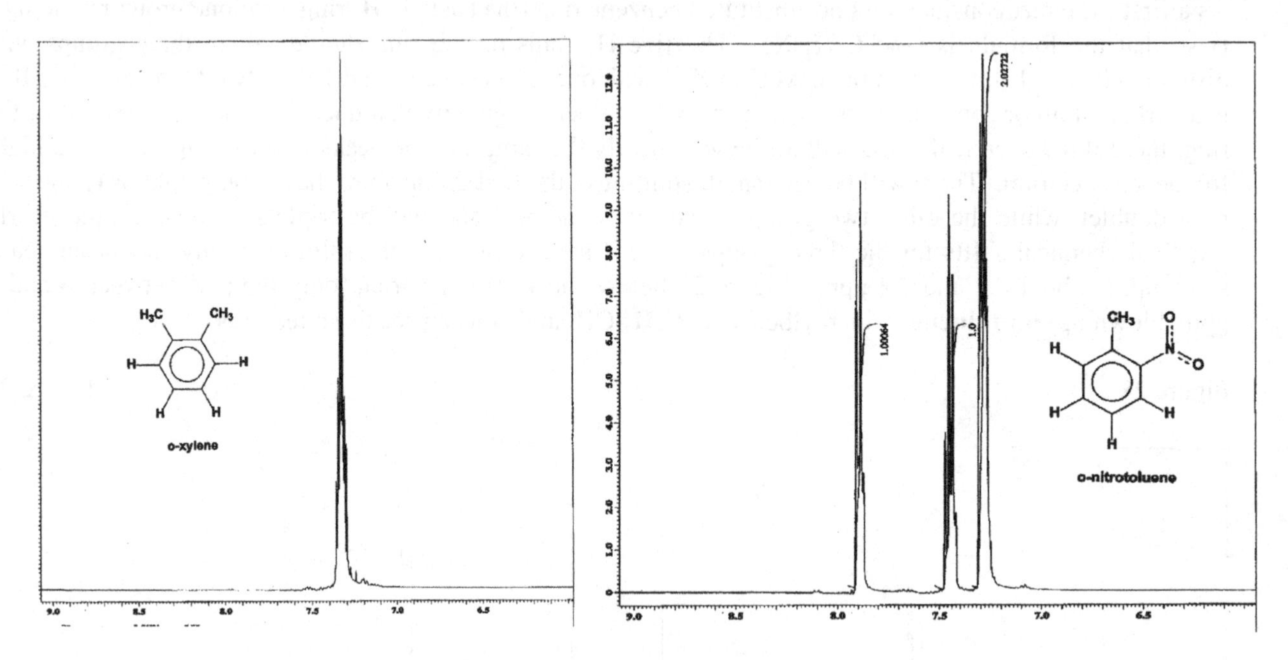

On the other hand, if the two groups are very different in electronegativity as in ***o*-nitrotoluene**, $CH_3C_6H_4$-NO_2, the four H atoms on the ring will all be symmetrically nonequivalent and we ought to see four different chemical shifts. In fact, Figure 31 shows this spectrum; there are three peaks, with areas 1:1:2. In this case, it's usually true that you can't separate out all four peaks because two of them will have nearly identical chemical shifts.

What changes if the two substituents in $\mathbf{YC_6H_4X}$ are *meta-* to each other (separated by one C atom on the ring)? If Y and X are the same or very similar, there should be three symmetrically distinct groups of H atoms: 1 H between X and Y, 2 next to X and Y on the outside, and 1 one carbon away from X and Y.

Figure 32 on the next page shows the 6-9 ppm region of the spectrum for ***m*-xylene**, $CH_3C_6H_4CH_3$. Only two peaks are seen, but the areas are 1:3 suggesting that two of the three peaks have nearly the same chemical shift. The smaller peak is a fairly clean triplet, suggesting that it's probably due to the third group above, and the larger peak must represent a singlet (first group) and doublet (second group) superimposed.

If we have a *meta-* geometry for YC_6H_4X but Y and X are very different, all four H atoms on the ring should be different. But in fact we see a spectrum much like the *ortho-* case: three peaks with areas 2:1:1, as in

Figure 32

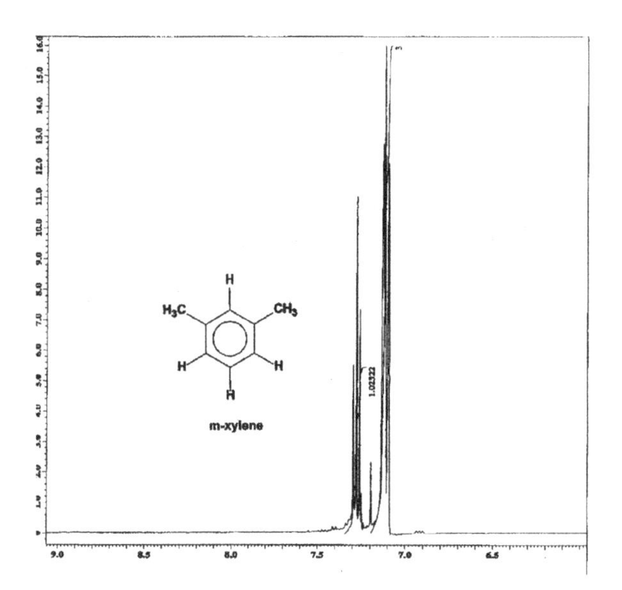

Figure 33

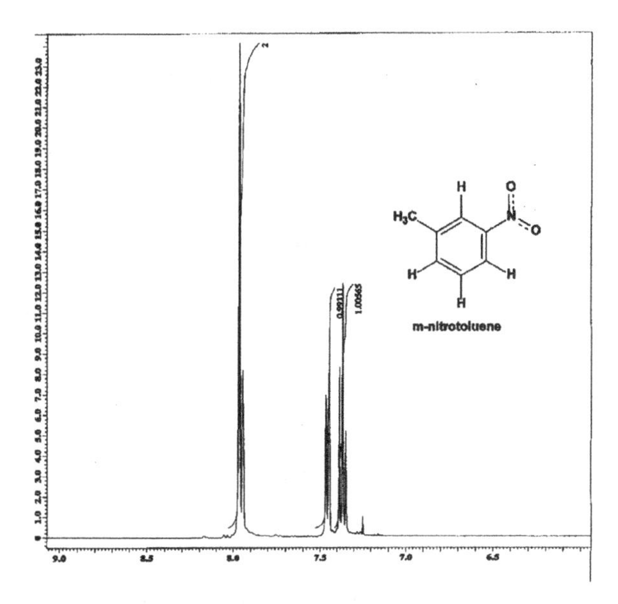

Figure 33 for ***m*-nitrotoluene**, $CH_3C_6H_4NO_2$ What's different from the *ortho-* spectrum is the relative chemical shift of the large peak. In the *meta-* case there are two H atoms next to the very electronegative NO_2 group, so that the area-2 peak gets shifted a lot (from about 7.2 to 8.0 ppm).

Now let's consider the *para-* geometry. This is very symmetrical; in fact, if X and Y in YC_6H_4X are the same, the four ring H atoms will be essentially the same and only one clean peak will result, as in Figure 34, which shows the aromatic region for ***p*-xylene**, $CH_3C_6H_4CH_3$. For this geometry, however, even a small difference between X and Y will lead to two peaks (2 H next to X, 2 H next to Y) with equal area, and in fact there will usually be a clean splitting of each peak into a doublet due to the single H neighbor. Figure 35 shows the aromatic region for ***p*-nitrotoluene**, $CH_3C_6H_4NO_2$, where the two substituents are very different and the peaks are fairly widely separated.

Figure 34

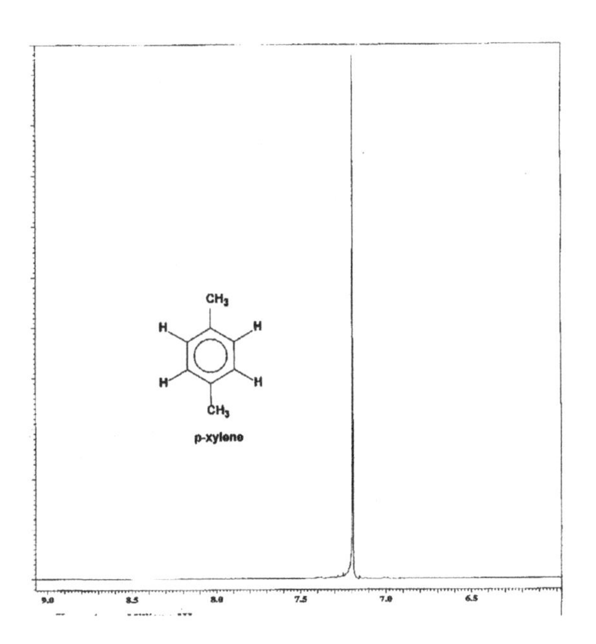

Figure 35

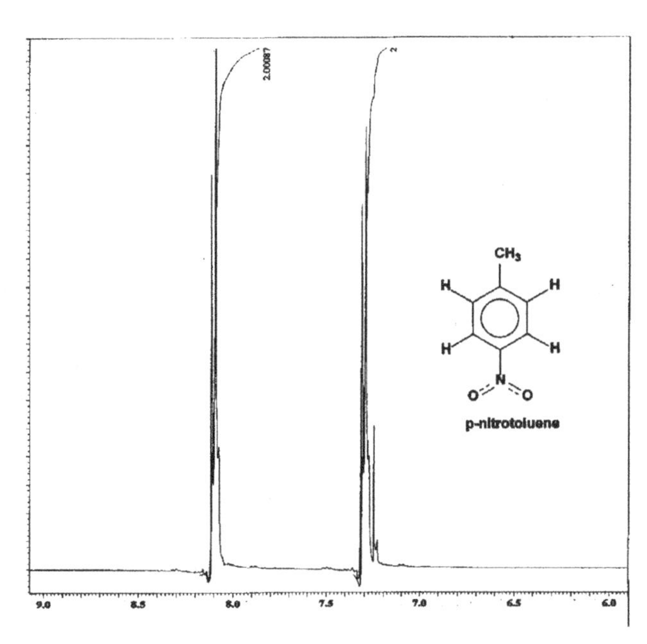

The bottom line for all of this discussion is that, given a high-resolution ^1H NMR spectrum, you can tell a lot about the geometrical arrangement of groups on a benzene ring by examining the aromatic region of the spectrum (between about 6.5 and 8.5 ppm) closely. This can be a big help for these molecules.

Sample Preparation

You can use the same samples for ^1H NMR that you made up for ^{13}C NMR earlier in the semester ***IF*** they haven't evaporated down. The capped tube should have sealed them adequately, but you need to check that the total liquid level is still the same as it was before. Your sample tubes should have been left upright in an Erlenmeyer flask in your desk, carefully capped. Remove them one at a time and check the level of each against the filling line on the NMR filling block on the lab bench. If the liquid level is within **five millimeters** of the filling line, just fill it to the mark with $CDCl_3$. If the liquid level is more than five millimeters below the filling line, you'll need to make up another sample: take the partially evaporated sample back to the general chem lab and make up a new sample, following the same procedure as before (page 87). In that case, dispose of the old sample by pouring it into the ***Waste $CDCl_3$ NMR samples*** bottle in the lab.

Running the spectrum and disposing of the samples

The instrument technician will insert the sample tube in the magnet for you and log you into the computer for the instrument. He'll walk you again through the menu on the screen, but you'll use the mouse to make choices with his help. Eventually you'll have two printed spectra, and the data will also be saved on the computer. Mount the spectra with tape in your notebook each on a separate left-hand page with the opposing right-hand page blank for your interpretation. Take the two color-coded sample tubes back to the lab, and pour each sample into the bottle in the hood marked "***Waste $CDCl_3$ NMR samples***". Don't rinse out the sample tubes, just put the original-color cap back on. Return the sample tubes immediately to the stockroom so they can be cleaned before stuff dries in them.

Notebook Writeup

This is a new experiment, of course; use the usual headings:

- **Purpose (step 14, page 30)**
- **Procedure (step 15)**
- **Apparatus (step 16 — include instrument: JEOL Eclipse+ 400 FT-NMR spectrometer)**
- **Data (step 17 — spectrum on LH page, table and interpretation RH page)**
- **Summary and Conclusions (step 20)**

When you get these spectra, the chemical shift values for the peaks present will be indicated under the peaks in *ppm*. Write this list down (in *ppm*, but only to two decimal places–1.13, not 1.1324) in your notebook on a RHP opposite the LHP spectrum *in tabular form* as ***Data***, labeled *a, b, c, ...* as indicated above (like your prediction tables). Include chemical shift, area, and splitting in your table. Look for the TMS reference peak, which should be exactly at 0.0000; if it's off, you might have to adjust your peak values appropriately. Compare this table, which is now extremely specific instead of being in general chemical-shift ranges, with your predictions that are already in your notebook. Note that the *a, b, c* sequence may be different from your prediction; what you called *d* in your prediction might be *b* here—that doesn't matter. What does matter is that you have the right number of peaks, with about the right chemical shifts, the right splittings, and the right relative areas. Describe your reasons for ruling out any structures that won't fit, and also give a specific fit for each of the hydrogens in the structure(or structures) that is/are compatible. Each possible structure needs at least a couple of lines of discussion. When you get at least one satisfactory spectral assignment (match of experiment with prediction), write up a short ***Summary*** about the status of your search and the existing candidates (with satisfactory boiling point range, ^{13}C NMR, IR, and ^1H NMR but focusing particularly on new information you got from ^1H NMR).

SUMMARY FOR LIQUID IDENTIFICATION

By the time you finish this project, you've done a number of experiments on your liquid mixture and written each experiment up separately. But there needs to be an overall summary that gives an all-inclusive view of what you've done and identifies the two liquids once and for all. Take a new right-hand page in your notebook and head it "***Liquid Unknown Summary***". Give a brief summary (half a page to a page) of the sequence of experiments you went through and how they gradually led you to an identification of each liquid. Be sure you *do* identify each liquid!! Once in a while somebody has shaky data and doesn't feel confident enough to make any identification at all—don't do that. It's always better to write down your best guess than to leave it blank. Even if you don't feel very confident, spend a few lines talking about what experimental evidence you relied on in picking the compound you did. **WRITE DOWN THE UNKNOWN NUMBER FROM THE ORIGINAL BOTTLE AND THE NAMES OF THE TWO COMPOUNDS YOU BELIEVE IT CONTAINED.** Having done this, prepare a table for each compound comparing the experimental value for the boiling point and the principal peaks for each molecular spectrum—^{13}C NMR, IR, and ^{1}H NMR—against the reference values for the compound you believe it to be. If it seems to you that there are significant discrepancies in the table, spend a little time discussing possible reasons for the discrepancies.

One final caution: More people miss identifying their compounds because they didn't sieve the table carefully enough than for any other reason. Be sure you have done a careful job of sieving for BP and matching candidates to their ^{13}C spectra before you come to a final decision on the identity of your compounds.

STRUCTURES, BOILING POINTS, AND BASIC SPECTROSCOPIC DATA FOR LIQUID UNKNOWN COMPOUNDS

The name of each compound is given as it appears in the *Aldrich Catalog Handbook*; when you have established candidates, you can look there for other physical data for each compound. Values are given from that catalog for the boiling point (in °C) for each compound. The column **C** gives the number of magnetically distinct carbon sites in the molecule—that is, the number of ^{13}C NMR peaks that should appear.

COMPOUND NAME	FORMULA	STRUCTURE	B.P.	C
Acetic anhydride	$CH_3COOCOCH_3$	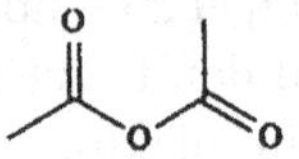	138	2
Acetone	CH_3COCH_3	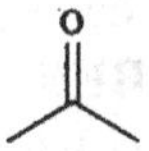	56	2
Acetonitrile	CH_3CN		82	2
Acetyl chloride	CH_3COCl	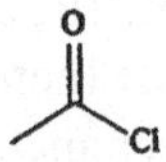	52	2
Acrylonitrile	$CH_2{=}CHCN$	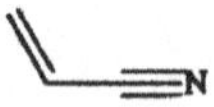	77	3
Allyl acetate	$CH_3COOCH_2CH{=}CH_2$	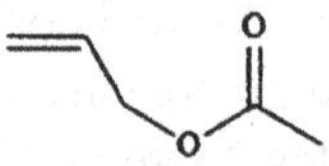	103	5
Allyl bromide	$CH_2{=}CHCH_2Br$		70	3
2-Aminoheptane	$CH_3CH(NH_2)(CH_2)_4CH_3$	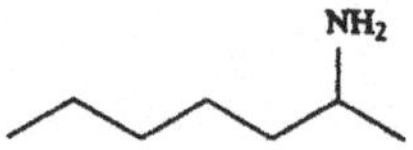	142	7
Amyl acetate	$CH_3COO(CH_2)_4CH_3$	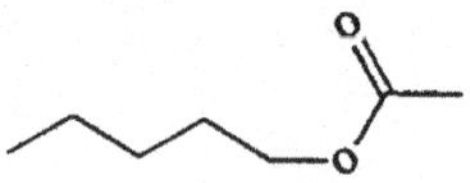	142	7
***tert*-Amyl alcohol**	$CH_3CH_2C(CH_3)_2OH$	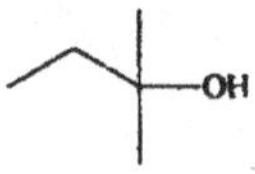	102	4
Anisole	$C_6H_5OCH_3$	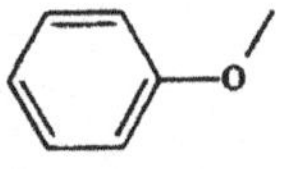	154	5
Benzaldehyde	C_6H_5CHO	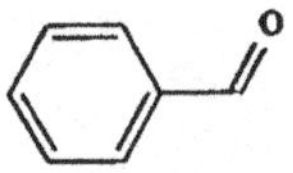	178	5
Benzene	C_6H_6		80	1
Bromobenzene	C_6H_5Br	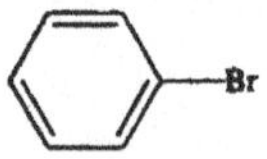	156	4
1-Bromobutane	$CH_3(CH_2)_3Br$		100	4
2-Bromobutane	$CH_3CHBrCH_2CH_3$	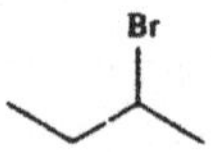	91	4
1-Bromohexane	$CH_3(CH_2)_5Br$		154	6
1-Bromopentane	$CH_3(CH_2)_4Br$	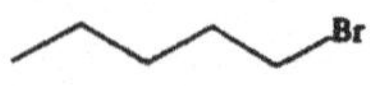	130	5
2-Bromopentane	$CH_3CH\ Br(CH_2)_2CH_3$	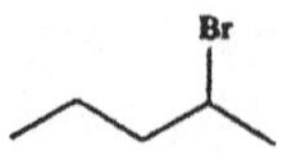	116	5

COMPOUND NAME	FORMULA	STRUCTURE	B.P.	C
1-Bromopropane	$CH_3CH_2CH_2Br$		71	3
2-Bromopropane	$(CH_3)_2CHBr$		59	2
1-Butanol	$CH_3(CH_2)_2CH_2OH$		118	4
2-Butanol	$CH_3CH(OH)CH_2CH_3$		99	4
2-Butanone	$CH_3COCH_2CH_3$		80	4
Butyl acetate	$CH_3COO(CH_2)_3CH_3$		124	6
tert-Butyl acetate	$CH_3COOC(CH_3)_3$		98	4
Butylamine	$CH_3(CH_2)_2CH_2NH_2$		78	4
Butyl ether	$[CH_3(CH_2)_3]_2O$		142	4
tert-Butyl methyl ether	$(CH_3)_3COCH_3$		53	3
Butyraldehyde	$CH_3(CH_2)_2CHO$		75	4
Butyric acid	$CH_3(CH_2)_2COOH$		162	4
Butyronitrile	$CH_3(CH_2)_2CN$		115	4
1-Chloro-2-methylpropane	$(CH_3)_2CHCH_2Cl$		68	3
2-Chloro-2-methylpropane	$(CH_3)_3CCl$		51	2
Chlorobenzene	C_6H_5Cl		132	4
1-Chlorobutane	$CH_3(CH_2)_2CH_2Cl$		77	4
Chloroform	$CHCl_3$		61	1
1-Chlorohexane	$CH_3(CH_2)_4CH_2Cl$		133	6
1-Chloropropane	$CH_3CH_2CH_2Cl$		46	3
2-Chlorotoluene	$ClC_6H_4CH_3$		157	7
3-Chlorotoluene	$ClC_6H_4CH_3$		160	7

COMPOUND NAME	FORMULA	STRUCTURE	B.P.	C
4-Chlorotoluene	$ClC_6H_4CH_3$	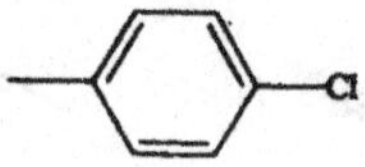	162	5
2,4,6-Collidine	$(CH_3)_3C_5H_2N$	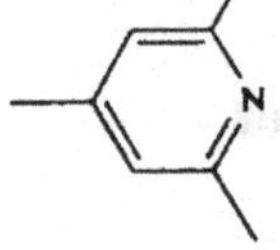	171	5
Crotonaldehyde	$CH_3CH=CHCHO$		104	4
Cumene	$C_6H_5CH(CH_3)_2$	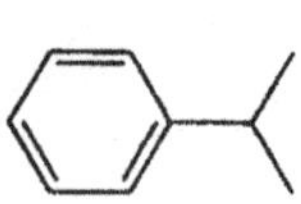	152	6
Cyclohexane	C_6H_{12}		81	1
Cyclohexanol	$C_6H_{11}OH$	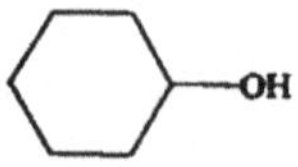	160	4
Cyclohexanone	$C_6H_{10}O$	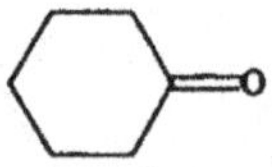	155	4
Cyclohexyl chloride	$C_6H_{11}Cl$	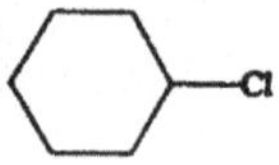	142	4
Cyclohexylamine	$C_6H_{11}NH_2$	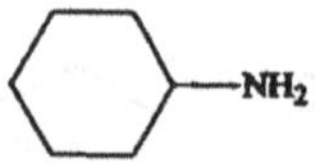	134	4
Cyclopentane	C_5H_{10}	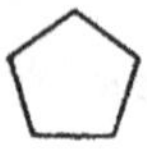	50	1
Cyclopentanol	C_5H_9OH	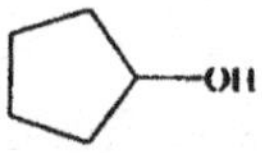	139	3
Cyclopentanone	C_5H_8O	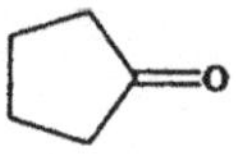	130	3
Cyclopentylamine	$C_5H_9NH_2$	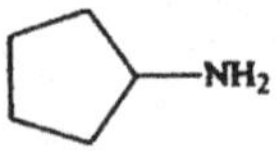	106	3
p-Cymene	$CH_3C_6H_4CH(CH_3)_2$	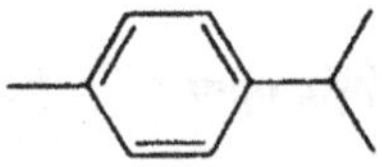	176	7
Decane	$CH_3(CH_2)_8CH_3$	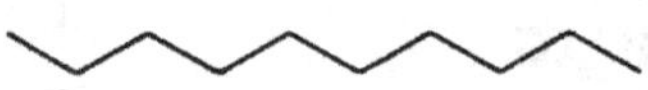	174	5
1,4-Diaminobutane	$H_2N(CH_2)_4NH_2$	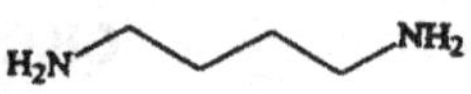	158	2
1,2-Diaminopropane	$H_2NCH_2CH(NH_2)CH_3$	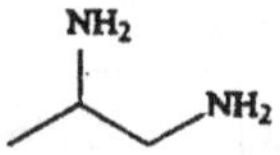	119	3
1,3-Diaminopropane	$H_2N(CH_2)_3NH_2$		140	2
1,2-Dibromoethylene	$BrCH=CHBr$		110	1
Dibromomethane	CH_2Br_2		96	1
Dibutylamine	$[CH_3(CH_2)_3]_2NH$	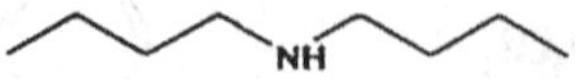	159	4
1,1-Dichloroethane	Cl_2CHCH_3	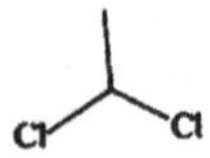	57	2

COMPOUND NAME	FORMULA	STRUCTURE	B.P.	C
1,2-Dichloroethane	$ClCH_2CH_2Cl$		83	1
1,2-Dichloroethylene	$ClCH=CHCl$	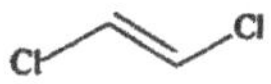	48	1
1,2-Dichloropropane	$ClCH_2CHClCH_3$	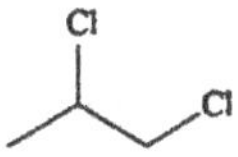	95	3
1,3-Dichloropropane	$(ClCH_2)_2CH_2$		120	2
Diethylamine	$(CH_3CH_2)_2NH$	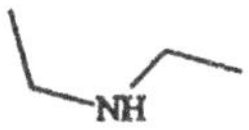	55	2
Diisopropylamine	$[(CH_3)_2CH]_2NH$	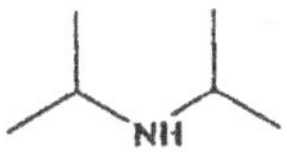	84	2
2,2-Dimethoxypropane	$(CH_3O)_2C(CH_3)_2$	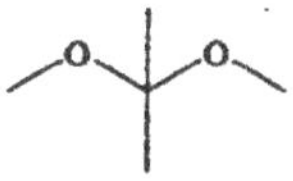	83	3
N,N-Dimethylacetamide	$CH_3CON(CH_3)_2$	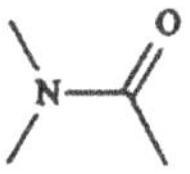	164	4
2,3-Dimethylbutane	$(CH_3)_2CHCH(CH_3)_2$	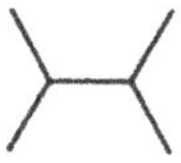	58	2
3,3-Dimethyl-1-butanol	$(CH_3)_3CCH_2CH_2OH$	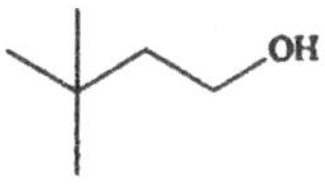	143	4
2,3-Dimethyl-2-butanol	$(CH_3)_2CHC(CH_3)_2OH$	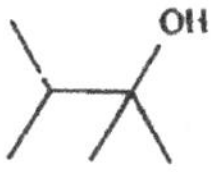	120	4
1,3-Dimethylbutylamine	$(CH_3)_2CHCH_2CH(CH_3)NH_2$	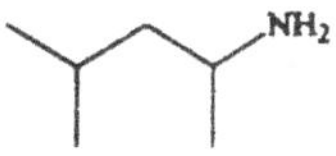	108	5
N,N-Dimethylformamide	$HCON(CH_3)_2$	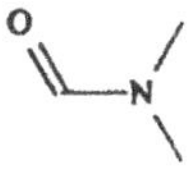	153	3
2,5-Dimethylhexane	$(CH_3)_2CH(CH_2)_2CH(CH_3)_2$	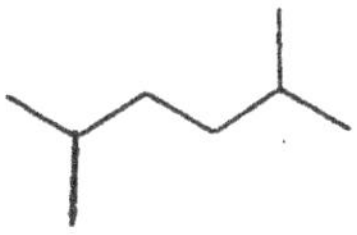	108	3
2,2-Dimethylhexane	$(CH_3)_3C(CH_2)_3CH_3$	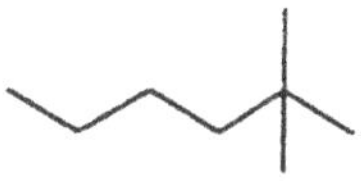	107	6
Dimethyl malonate	$CH_3OOCCH_2COOCH_3$	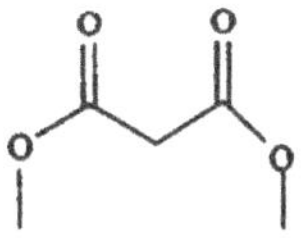	180	3
3,3-Dimethylpentane	$(CH_3CH_2)_2C(CH_3)_2$	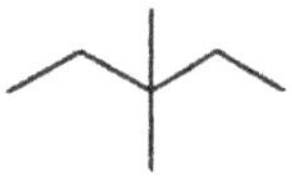	86	4
2,3-Dimethylpentane	$(CH_3)_2CHCH(CH_3)CH_2CH_3$	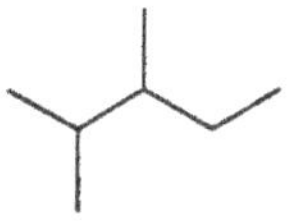	89	6
2,2-Dimethylpentane	$(CH_3)_3C(CH_2)_2CH_3$	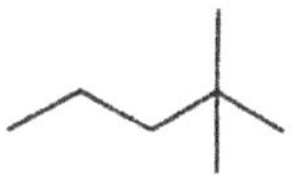	78	5
2,4-Dimethyl-3-pentanol	$[(CH_3)_2CH]_2CHOH$	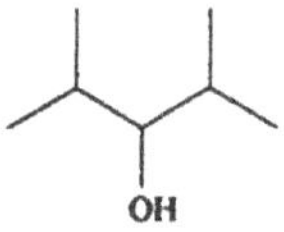	139	4
2,4-Dimethyl-3-pentanone	$[(CH_3)_2CH]_2CO$	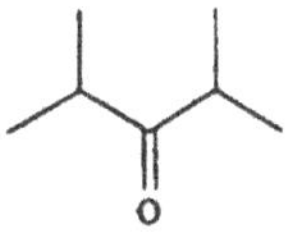	124	3
4,4-Dimethyl-2-pentanone	$CH_3COCH_2C(CH_3)_3$	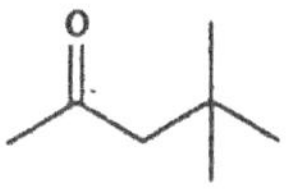	125	5

COMPOUND NAME	FORMULA	STRUCTURE	B.P.	C
1,4-Dioxane	$C_4H_8O_2$	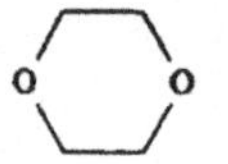	101	1
Dipropylamine	$(CH_3CH_2CH_2)_2NH$	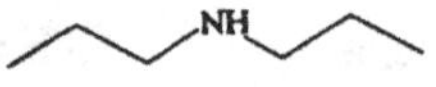	105	3
2-Ethoxyethanol	$CH_3CH_2OCH_2CH_2OH$	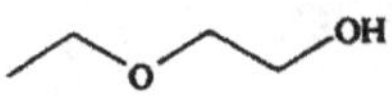	135	4
Ethyl acetate	$CH_3COOCH_2CH_3$	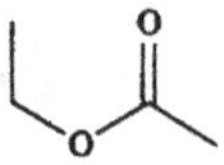	77	4
Ethyl alcohol (ethanol)	CH_3CH_2OH	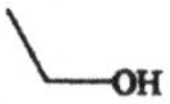	78	2
Ethylbenzene	$C_6H_5CH_2CH_3$	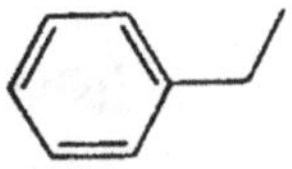	136	6
2-Ethyl-1-butanol	$(CH_3CH_2)_2CHCH_2OH$	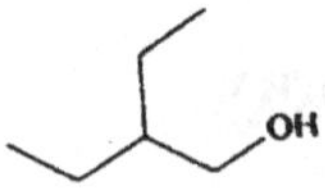	146	4
2-Ethylbutyraldehyde	$(CH_3CH_2)_2CHCHO$	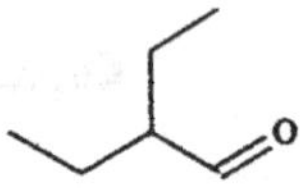	117	4
Ethyl butyrate	$CH_3(CH_2)_2COOCH_2CH_3$	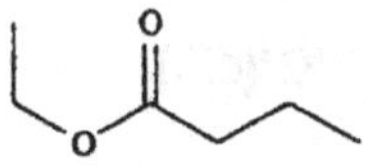	120	6
Ethyl crotonate	$CH_3CH{=}CHCOOCH_2CH_3$	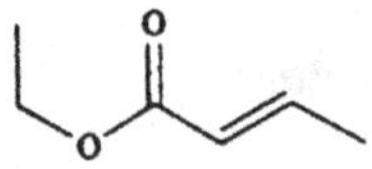	142	6
Ethylene glycol diethyl ether	$CH_3CH_2OCH_2CH_2OCH_2CH_3$	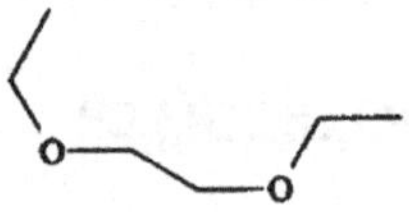	121	3
Ethylene glycol dimethyl ether	$CH_3OCH_2CH_2OCH_3$		85	2
Ethyl isobutyrate	$(CH_3)_2CHCOOCH_2CH_3$	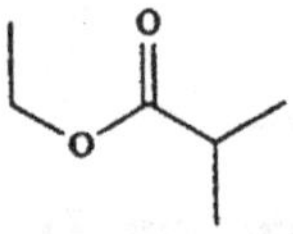	112	5
Ethyl isovalerate	$(CH_3)_2CHCH_2COOCH_2CH_3$	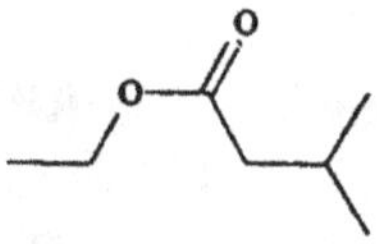	131	6
Ethyl propionate	$CH_3CH_2COOCH_2CH_3$	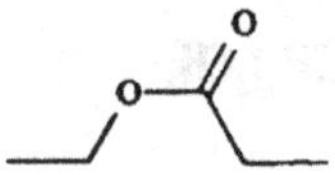	99	5
Ethyl pyruvate	$CH_3COCOOCH_2CH_3$	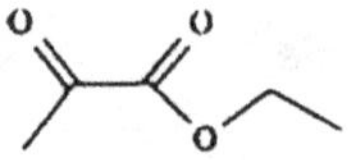	144	5
2-Ethyltoluene	$CH_3C_6H_4CH_2CH_3$	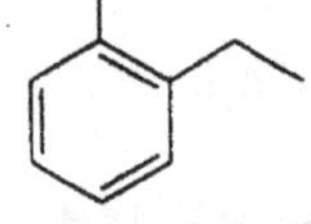	164	9
3-Ethyltoluene	$CH_3C_6H_4CH_2CH_3$	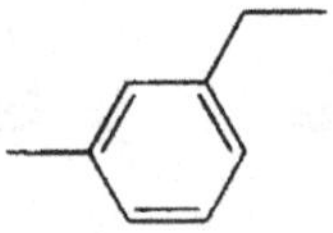	158	9
4-Ethyltoluene	$CH_3C_6H_4CH_2CH_3$	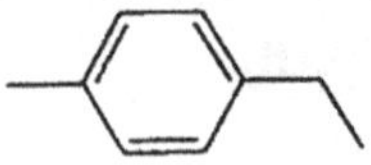	162	7
Heptaldehyde	$CH_3(CH_2)_5CHO$	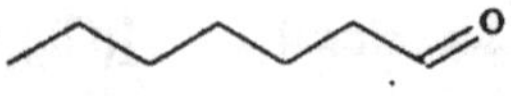	153	7
Heptane	$CH_3(CH_2)_5CH_3$		98	4
1-Heptanol	$CH_3(CH_2)_5CH_2OH$		176	7

COMPOUND NAME	FORMULA	STRUCTURE	B.P.	C
2-Heptanol	$CH_3CH(OH)(CH_2)_4CH_3$		160	7
2-Heptanone	$CH_3CO(CH_2)_4CH_3$		149	7
3-Heptanone	$CH_3CH_2CO(CH_2)_3CH_3$	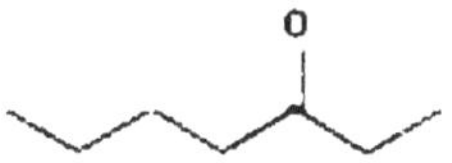	146	7
4-Heptanone	$(CH_3CH_2CH_2)_2CO$		145	4
1-Heptene	$CH_2=CH(CH_2)_4CH_3$		94	7
1,5-Hexadiene	$CH_2=CH(CH_2)_2CH=CH_2$		60	3
2,4-Hexadiene	$CH_3CH=CHCH=CHCH_3$		82	3
Hexanal	$CH_3(CH_2)_4CHO$		131	6
Hexane	$CH_3(CH_2)_4CH_3$		68	3
2-Hexanol	$CH_3CH(OH)(CH_2)_3CH_3$	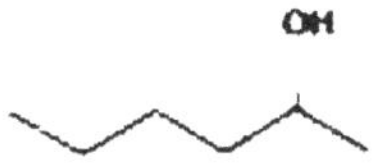	136	6
3-Hexanol	$CH_3CH_2CH(OH)(CH_2)_2CH_3$	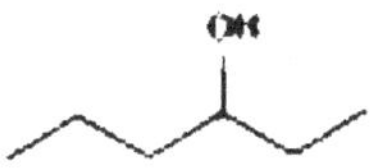	135	6
2-Hexanone	$CH_3CO(CH_2)_3CH_3$	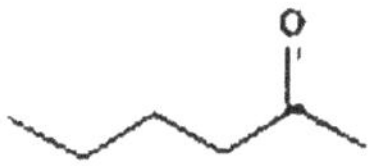	127	6
3-Hexanone	$CH_3CH_2CO(CH_2)_2CH_3$	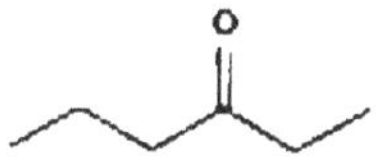	123	6
1-Hexene	$CH_2=CH(CH_2)_3CH_3$		60	6
Hexyl acetate	$CH_3COO(CH_2)_5CH_3$		168	8
Hexyl alcohol	$CH_3(CH_2)_4CH_2OH$		157	6
Hexylamine	$CH_3(CH_2)_4CH_2NH_2$		131	6
1-Iodobutane	$CH_3CH_2CH_2CH_2I$		130	4
1-Iodopropane	$CH_3CH_2CH_2I$		101	3
2-Iodopropane	$(CH_3)_2CH\ I$		88	2
Isoamyl acetate	$CH_3COO(CH_2)_2CH(CH_3)_2$		142	6
Isoamylamine	$(CH_3)_2CHCH_2CH_2NH_2$		95	4

COMPOUND NAME	FORMULA	STRUCTURE	B.P.	C
Isobutyl acetate	$CH_3COOCH_2CH(CH_3)_2$	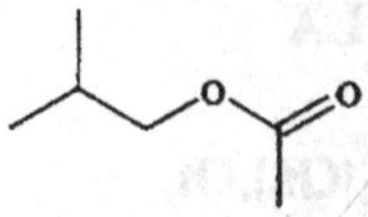	115	5
Isobutyraldehyde	$(CH_3)_2CHCHO$	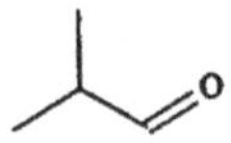	63	3
Isopropyl acetate	$CH_3COOCH(CH_3)_2$	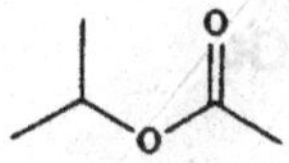	85	4
Isopropyl ether	$[(CH_3)_2CH]_2O$	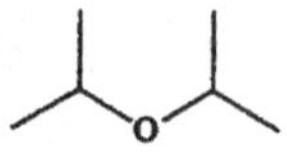	68	2
Isovaleraldehyde	$(CH_3)_2CHCH_2CHO$	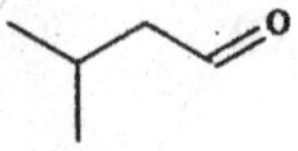	90	4
Isovaleric acid	$(CH_3)_2CHCH_2COOH$	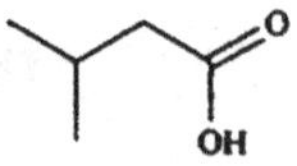	175	4
2,4-Lutidine	$(CH_3)_2C_5H_3N$	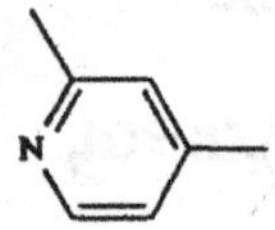	159	7
2,5-Lutidine	$(CH_3)_2C_5H_3N$	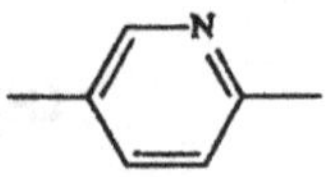	157	7
2,6-Lutidine	$(CH_3)_2C_5H_3N$	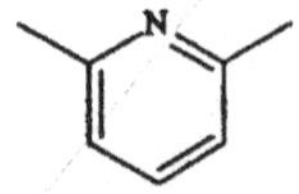	143	4
3,4-Lutidine	$(CH_3)_2C_5H_3N$	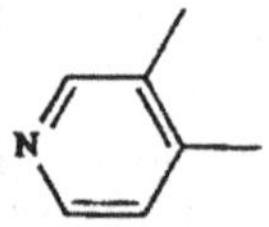	163	7
2-Methoxyethanol	$CH_3OCH_2CH_2OH$		124	3
2-Methoxyethyl ether	$[CH_3OCH_2CH_2]_2O$	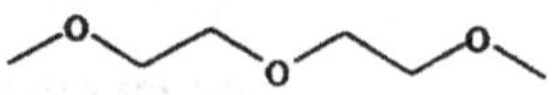	162	3
Methyl acetate	CH_3COOCH_3	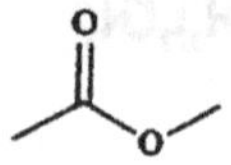	57	3
Methyl alcohol (methanol)	CH_3OH		65	1
2-Methyl-1-butanol	$CH_3CH_2CH(CH_3)CH_2OH$	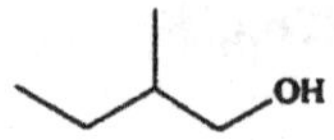	130	5
3-Methyl-1-butanol	$(CH_3)_2CHCH_2CH_2OH$	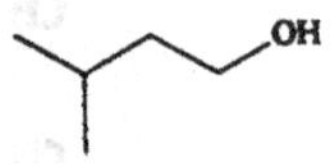	130	4
3-Methyl-2-butanone	$(CH_3)_2CHCOCH_3$	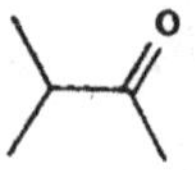	94	4
Methyl butyrate	$CH_3(CH_2)_2COOCH_3$	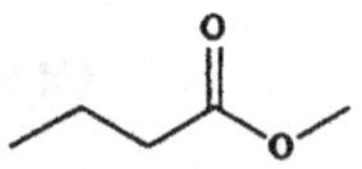	102	5
Methylcyclohexane	$C_6H_{11}CH_3$	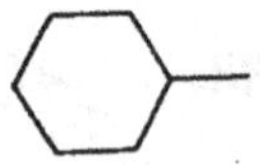	101	5
1-Methylcyclohexanol	$C_6H_{10}(CH_3)OH$		168	5
Methylcyclopentane	$C_5H_9CH_3$	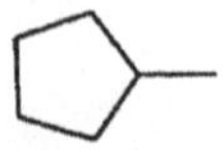	72	4
1-Methyl-1-cyclopentene	$C_5H_7CH_3$	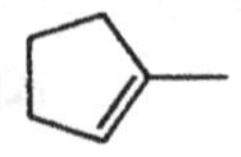	72	6

COMPOUND NAME	FORMULA	STRUCTURE	B.P.	C
2-Methylheptane	$(CH_3)_2CH(CH_2)_4CH_3$		116	7
3-Methylhexane	$CH_3CH_2CH(CH_3)(CH_2)_2CH_3$	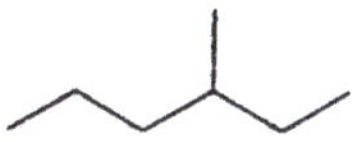	91	7
Methyl isobutyrate	$(CH_3)_2CHCOOCH_3$	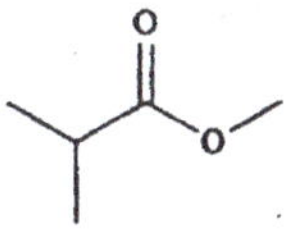	90	4
2-Methylpentane	$(CH_3)_2CH(CH_2)_2CH_3$	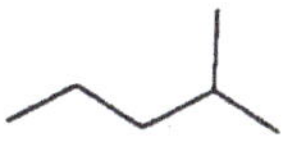	62	5
3-Methylpentane	$(CH_3CH_2)_2CHCH_3$		64	4
3-Methyl-1-pentanol	$CH_3CH_2CH(CH_3)CH_2CH_2OH$	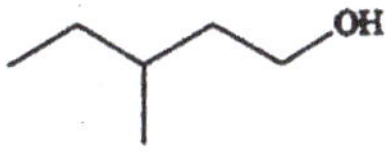	151	6
3-Methyl-2-pentanol	$CH_3CH_2CH(CH_3)CH(OH)CH_3$	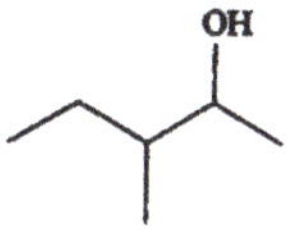	131	6
3-Methyl-3-pentanol	$(CH_3CH_2)_2C(CH_3)OH$	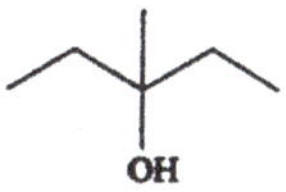	123	4
4-Methyl-2-pentanol	$(CH_3)_2CHCH_2CH(OH)CH_3$	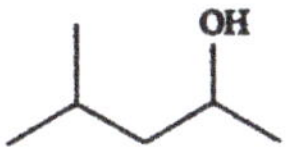	132	6
3-Methyl-2-pentanone	$CH_3CH_2CH(CH_3)COCH_3$	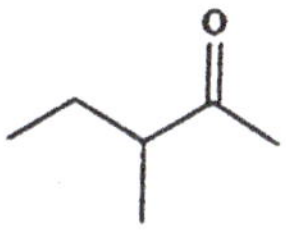	118	6
4-Methyl-2-pentanone	$(CH_3)_2CHCH_2COCH_3$	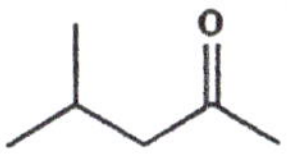	116	5
2-Methyl-1-propanol	$(CH_3)_2CHCH_2OH$	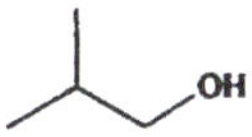	108	3
2-Methyl-2-propanol	$(CH_3)_3COH$	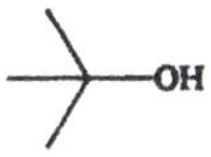	83	2
Methyl propionate	$CH_3CH_2COOCH_3$	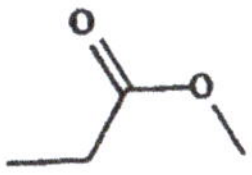	79	4
Methyl trimethylacetate	$(CH_3)_3CCOOCH_3$	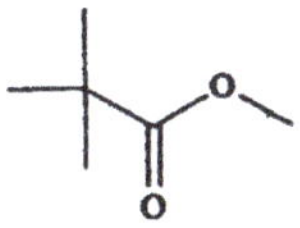	101	4
Methyl valerate	$CH_3(CH_2)_3COOCH_3$	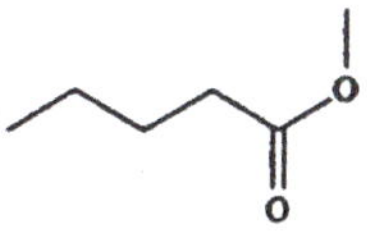	128	6
Nitroethane	$CH_3CH_2NO_2$	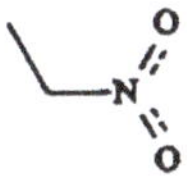	112	2
Nitromethane	CH_3NO_2	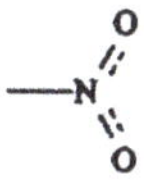	100	1
1-Nitropropane	$CH_3CH_2CH_2NO_2$	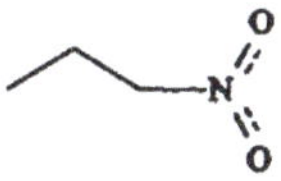	131	3
2-Nitropropane	$(CH_3)_2CHNO_2$	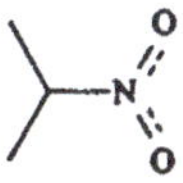	120	2
Nonane	$CH_3(CH_2)_7CH_3$		151	5
Octane	$CH_3(CH_2)_6CH_3$		125	4

COMPOUND NAME	FORMULA	STRUCTURE	B.P.	C
2-Octanol	$CH_3(CH_2)_5CH(OH)CH_3$	OH	175	8
2-Octanone	$CH_3(CH_2)_5COCH_3$	O	173	8
3-Octanone	$CH_3(CH_2)_4COCH_2CH_3$	O	167	8
1-Pentanol	$CH_3(CH_2)_3CH_2OH$	OH	136	5
2-Pentanol	$CH_3(CH_2)_2CH(OH)CH_3$	OH	118	5
3-Pentanol	$(CH_3CH_2)_2CHOH$	OH	114	3
2-Pentanone	$CH_3(CH_2)_2COCH_3$	O	100	5
3-Pentanone	$(CH_3CH_2)_2CO$	O	102	3
3-Picoline	$CH_3C_5H_4N$	N	143	6
4-Picoline	$CH_3C_5H_4N$	N	145	4
1-Propanol	$CH_3CH_2CH_2OH$	OH	97	3
2-Propanol	$(CH_3)_2CHOH$	OH	82	2
Propionaldehyde	CH_3CH_2CHO	O	46	3
Propionic acid	CH_3CH_2COOH	OH, O	141	3
Propionitrile	CH_3CH_2CN	N	97	3
Propyl acetate	$CH_3COOCH_2CH_2CH_3$	O, O	102	5
Propylamine	$CH_3CH_2CH_2NH_2$	NH_2	48	3
Propyl ether	$(CH_3CH_2CH_2)_2O$	O	88	3
Pyridine	C_5H_5N	N	115	3
Styrene	$C_6H_5CH=CH_2$		146	6
Tetrachloroethylene	$Cl_2C=CCl_2$	Cl, Cl, Cl, Cl	121	1
Tetrahydrofuran	C_4H_8O	O	67	2

COMPOUND NAME	FORMULA	STRUCTURE	B.P.	C
Toluene	$C_6H_5CH_3$	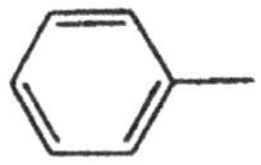	111	5
1,1,1-Trichloroethane	Cl_3CCH_3	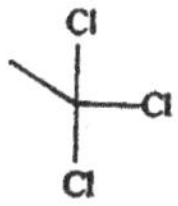	74	2
Trichloroethylene	$Cl_2C=CHCl$	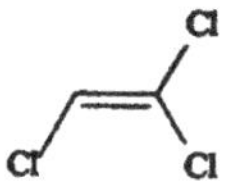	87	2
1,2,3-Trichloropropane	$(CH_2Cl)_2CHCl$	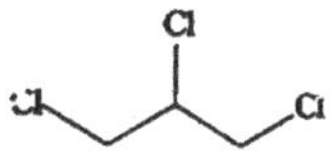	156	2
Triethylamine	$(CH_3CH_2)_3N$	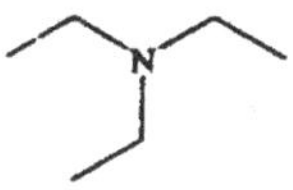	89	2
1,2,3-Trimethylbenzene	$C_6H_3(CH_3)_3$	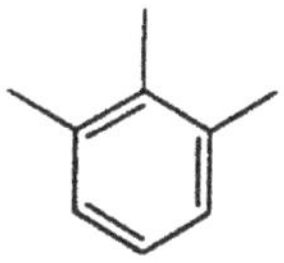	175	6
1,2,4-Trimethylbenzene	$C_6H_3(CH_3)_3$	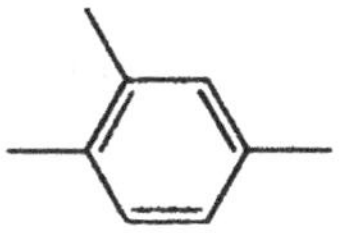	168	9
2,2,3-Trimethylbutane	$(CH_3)_2CHC(CH_3)_3$		81	4
2,2,4-Trimethylpentane	$(CH_3)_2CHCH_2C(CH_3)_3$	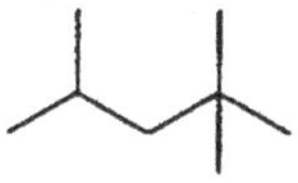	98	5
Tripropylamine	$(CH_3CH_2CH_2)_3N$	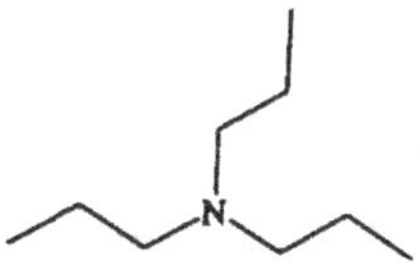	155	3
Valeronitrile	$CH_3(CH_2)_3CN$		139	5
Vinyl acetate	$CH_3COOCH=CH_2$	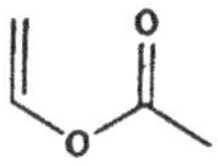	72	4
o-Xylene	$C_6H_4(CH_3)_2$	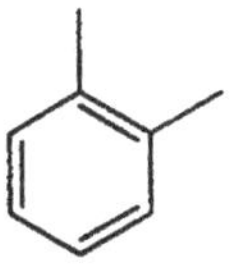	143	4
m-Xylene	$C_6H_4(CH_3)_2$	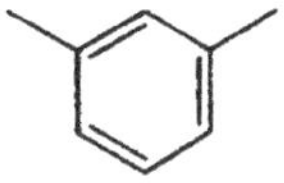	138	5
p-Xylene	$C_6H_4(CH_3)_2$		138	3

IDENTIFICATION OF SOLID ORGANIC ACIDS

You will determine the melting point (MP), neutralization equivalent (NE), acid ionization constant (pK_a), and ^1H NMR spectrum of your unknown acid. Since neutralization equivalent is to be calculated to 4 significant figures in order to obtain 3 good significant figures, it will be necessary for you to exercise considerable care in order to obtain adequate precision; this is analytical chemistry that needs precision comparable to the titrations in analyzing your coordination compound in Chem 151.

MELTING POINT (3 separate runs, each about 20 min; start on notebook p. 55)

Just as the boiling point is a convenient, valuable numerical characterization for a liquid, the melting point is similarly valuable for a solid. In fact, the vast majority of pharmaceuticals are solids and melting point is so valuable as an indication of purity that the U.S. Pharmacopeia gives strict rules for melting point procedures (it's a multi-billion-dollar industry).

The acid you receive is from the stockroom supply. No impurities have been added, but whatever the manufacturer left there is still there. So we're going to want you to check the melting point of your solid acid, purify it by recrystallization, and get a better, more accurate, value of the melting point on the recrystallized acid. The purer the acid, the higher its melting point will be (remember from colligative properties that impurities lower the freezing point – same thing as melting point). Just as for liquids and boiling point, there will be a melting point *range*, and the range will be smaller the more nearly pure the solid is. A reasonably pure solid compound will have a melting point range of only about 1°C, but to get that level of accuracy you'll have to heat it very slowly to be sure everything's at thermal equilibrium.

The slow heating is a problem when you don't have any idea what the melting point is: heating the solid sample at 1°/minute from 50° to 350° would take five hours, obviously. Not good. So the thing to do is to run a quickie at a heating rate of 10°/min from about 60° to 260°, which is high enough to catch all of the acids on our list. That'll take 20 minutes, but fortunately three of you can do it at once. The result won't really be a melting point, but it *will* give you a short range of temperature that the melting point range lies in. Then you can use the apparatus (usually called a ***melting point block***) by yourself, at 1°/min, over a short range around your melting point, to get a first pass at the true melting point range. After you recrystallize and dry your solid acid, you can get a better value of the melting point range by checking the recrystallized material at 0.5°/min.

Get a clean melting-point capillary from the vial on the desk with the MP blocks or from the instructor. Fill it about 2 mm deep – that's less than an eighth of an inch – with powdered acid. That's a teeny little hole, so how do you get the solid inside? Pour a *small* amount of the acid out on a clean smooth watch glass and press the open end of the capillary down into the solid. Two or three presses should get you a 2-mm plug inside the open end. Now drop the capillary, closed end on the bottom, down through a 3-foot piece of glass tubing held against the bench top or floor. The jolt at the bottom should knock your sample to the bottom of the capillary. Clean the capillary off by wiping it with a Kimwipe.

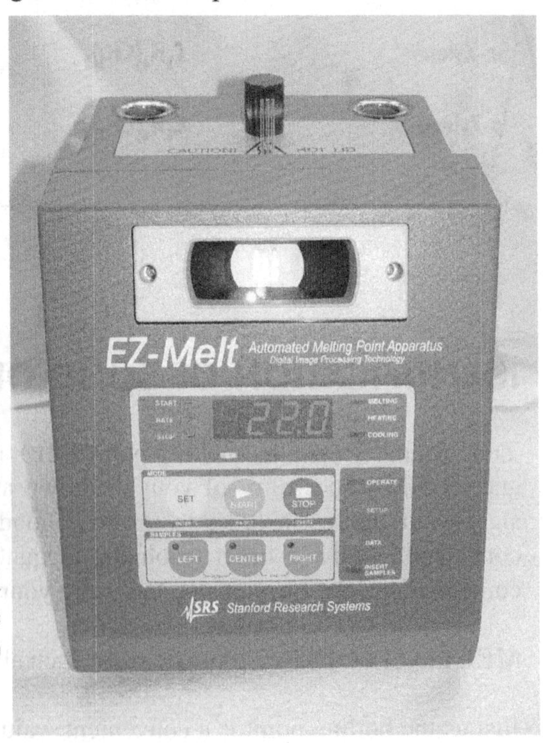

Now get two other guys with capillaries to run the fast pass at the same time. Start a new notebook page with Purpose, Procedure, etc. *Don't* put the capillaries in yet.

- Be sure it's turned on (switch at center of bottom back side).
- Press the yellow **SET** button once to display the start temp. The yellow **SETUP** LED turns on to show that the temp reading is *not* the current oven temp; the yellow **START** LED indicates that the display is reading what will be the starting temp for the temperature ramp. Use RAISE and LOWER buttons (labels for the green **START** and red **STOP** buttons) to set the start temp to 60.0°C.
- Press the **SET** button again to accept the start temp and move to choosing the temperature ramp heating rate. The **RATE** LED will light. Choose a ramp rate of 10 degrees/minute with the RAISE and LOWER buttons.

- Press the **SET** button again to accept the ramp rate and move to choosing the stop temperature (the upper limit of the ramp). The **STOP** LED will light. Use the RAISE and LOWER buttons to set a stop temperature of 260.0°C. Press the **SET** button again to accept the upper limit. The **SETUP** LED should turn off and the display should show the current oven temp.

- Press the green **START** button to begin preheating to the start temp; the **HEATING** LED will turn on. The block will beep when the start temp has been reached, in only a couple of minutes, and the **INSERT SAMPLES** LED will light.

- Each of the three of you should drop his capillary *vertically* straight down into one of the three small holes in the top front (see figure). It should drop in freely; don't push the capillary! The display will show the oven cooling off due to the cool capillaries; wait a few seconds for it to stabilize. Use the time to record in your notebook which sample position you used (left, center, right) and who has samples in the other two holes.

- Press the **START** button again to begin the 10°/min ramp. Watch your capillary through the magnifying window. You're looking for two events: (1) The sample turns soggy inside the capillary and starts to make little wet spots against the glass – this is called the ***onset point***. (2) The soggy sample melts completely to a clear liquid with no particles left – this is the ***clear point***. Record the temperature for each, as you observe it, in a data table in your notebook as shown below (but note the next two steps).

	Observed	Auto
Onset point (°C)		
Clear point (°C)		

- When all three samples have melted, press the red **STOP** button to end the ramp. This might be long before the temperature reaches the upper limit. The **DATA** LED should come on blinking. This means the MP block has memorized the two temperatures you recorded, by using a digital camera image, for each sample.
- If you had the left sample, press the blue **LEFT** button to display the block's version of the onset point; record that as *auto onset point* in your notebook. Press **LEFT** again to display the block's version of the clear point; record that as *auto clear point* in your notebook. Then the student with the center sample can do the same thing with the **CENTER** button, and the student with the right-hand sample can do the same thing with the **RIGHT** button.

- Press **STOP** to end the run. *Carefully* remove your capillary from its hole by lifting *vertically* – don't break it off! Put the used capillary in one of the two receiver tubes at the back of the block. Leave the power on; the instrument will cool automatically.

Now look at the temperatures you've recorded: yours and the block's. If they're reasonably near each other, assume the block is right and go with the "Auto" temperatures. If they're wildly different, ask one of us about the discrepancy. What you have at this point is a really awful version of the melting point range – but what it can do is to guide you to a convenient determination of the *true* MP range. What you will need to do is to use the block alone and program it for the temperatures you've observed, but go through it slowly. Fast is bad, or at least crude: consider the table on the next page (taken from the block's manual, for a standard compound).

Ramp Rate (°C/min)	Clear Point (°C)	MP Error (°C)
0.1	134.2	0
0.2	134.4	0.2
0.5	134.9	0.7
1	135.4	1.2
2	136.2	2
5	137.9	3.7

The error at 10°/min is so bad they didn't put it in the table. It's probably high by somewhere around 5°C. So, next (though this could be another week in lab, whenever it's convenient), get a MP block to yourself. Go through the programming steps at the beginning of the bulleted list above, setting the start temp at about 6° below the onset point you observed in the fast run. Set the ramp at 1°C/min. Set the upper limit (the stop temp) at 4° above the clear point – the upper temp – you observed in the fast run. This should be a total range of maybe 12-15 degrees, or 12-15 minutes. While the block is heating up, make up a *new* capillary sample and wipe it clean. When the block beeps that it's at the start temp, insert the capillary and press the **START** button; carry on from there to the end of the procedure. The two temperatures you record this time should be lower and closer together. They probably represent a pretty good version of the true MP range. Record them *and* record the ramp rate of 1°/min.

Once you've recrystallized the solid acid and dried it (over good desiccant) for a week, run a MP again on the recrystallized solid, programming for a ramp rate of 0.5°/min, a start temp about 2° below the onset point from your unrecrystallized solid MP range, and a stop temp about 2° above the clear point from that range. Record the two temps and the ramp rate. This should be a reliable version of the melting point of your acid.

Notebook Writeup
The melting-point determination is a new experiment, of course; use the usual headings:

- **Purpose (step 14, page 31)**
- **Procedure (step 15)**
- **Apparatus (step 16 — include instrument: SRS MP block)**
- **Data (step 17 — temperature ranges)**
- **Summary and Conclusions (step 20)**

RECRYSTALLIZATION (about 90 minutes; start on notebook page 59)

Nearly all solids exist as crystals, particularly reasonably simple organic and inorganic solid compounds. When crystalline compounds contain impurities, they can often be effectively removed by dissolving the crystals in the *minimum possible* amount of a hot solvent and re-forming them out of that cooling solution, in which the good stuff is quite concentrated but the impurities are fairly dilute.

What you need for recrystallization purposes is a solvent that dissolves your acid hot but not cold. If you haven't already checked this, try a pinch of your acid in about a mL of water in a small test tube at room temperature, then heat it if it *doesn't* dissolve to see if it goes in hot. If it dissolves in water cold, that's too soluble—go on to ethanol at room temperature. If it won't go in cold ethanol, heat it in a beaker of hot water (not with a burner). Pick the solvent that the acid is *least* soluble in cold.

Put *half* your acid unknown in a clean dry beaker (at least 250-mL). Estimate roughly the amount of solvent that will be needed to dissolve the acid hot, and place that amount in another clean beaker. Heat the solvent to boiling on a hot plate in the hood. Pour a small amount of boiling solvent over the solid acid and stir it with a clean stirring rod or a wood stick to see if it goes into solution. If it does, stop! If it doesn't dissolve completely, put both beakers on the hot plate and keep adding boiling solvent *a little at a time* till all the solid has dissolved (over a period of several minutes). Leave the wood stick in the beaker while heating. When you have added *just enough boiling solvent* to dissolve the solid, stop. Take both beakers off the hot plate and let the solution stand for about 30 minutes to form crystals. Then place the beaker with crystals in an ice bath for another 20-30 minutes. When it's thoroughly cold and crystals have stopped forming, filter the crystals off using your suction flask and Buchner funnel. Most acids will dry readily by simply transferring the crystals to an evaporating dish and leaving them in your desiccator (covered by a ribbed watch glass) for a week. Sometimes a vacuum desiccator is needed; ask an instructor if you should do this. When the acid crystals are thoroughly dry, check the melting point again. Compare the new value to the original value, both in terms of temperature values and range.

Notebook Writeup

Recrystallization is also a new experiment; use the usual headings in your notebook (but add the experiment title both in your notebook and in the table of contents):

- **Purpose (step 14, page 31)**
- **Procedure (step 15)**
- **Apparatus (step 16)**
- **Data (step 17 — observations as in synthesis, wt. yield when dried)**
- **Summary and Conclusions (step 20)**

NEUTRALIZATION EQUIVALENT AND pK_a (about one lab period; start on notebook page 65)

Another very important property of acids, this time a chemical property, is their neutralization equivalent. An acid's ***neutralization equivalent*** is defined as the number of grams of the acid which will yield one gram-atomic weight of H_3O^+, or, alternatively, as the number of grams of the acid which will react with one formula weight of NaOH. If the acid has only one transferrable proton, the neutralization equivalent will be equal to the molecular weight, but if it has two transferrable protons, the neutralization equivalent will be half the molecular weight (because it only takes half a mole of molecules to yield a whole mole of protons), and so on. The reason this concept of neutralization equivalent is necessary is that the determination you will make involves a known number of formula weights of NaOH but only a known weight of acid, so that the only quantity which can be determined is the neutralization equivalent defined as above. The number of *moles* of acid can't be determined by this method unless the number of replaceable protons is known.

The determination of the neutralization equivalent rests on the fact that there is a reaction which all of these acids will undergo completely, namely, that with OH^- ion:

$$\text{R-COOH} + OH^- \rightarrow \text{R-COO}^- + H_2O$$

This can obviously be turned into a titration – make up a solution of, say, NaOH that has a known molar concentration, put that solution into a buret, and titrate an accurately weighed sample of the acid (dissolved in water in the flask). We need an indicator that changes color when the acid is just used up, so that the color change at the end-point will mean that the number of moles of base added is equal at that point to the number of moles of acid originally present:

$$\text{moles acid} = \text{moles base}$$

$$\frac{\text{grams acid}}{\text{g acid}/\text{mole acid}} = \text{liters base} \times \frac{\text{moles base}}{\text{liter}}$$

or $$\text{acid sample wt} / \text{NE} = \text{buret reading at endpoint} \times \text{molarity of NaOH}$$

Then $$\text{NE} = \frac{\text{acid sample wt}}{\text{buret rdg in L} \times \text{M(NaOH)}}$$

How can we be certain of the concentration of the OH^- solution (or of the titrant in any titration)? A few compounds have sufficiently precise stoichiometry and are sufficiently stable to humid air for it to be possible to weigh them out directly and dissolve the weighed sample in water to be diluted in a volumetric flask to a prescribed volume; these substances are called ***primary standards*** (see p. 17). Unfortunately NaOH is not one of these, since it absorbs water and CO_2 from the air quite rapidly and thus can't be accurately weighed. In cases such as this the solution must be made up with approximately the desired concentration and standardized against some primary standard, just as you standardized your EDTA solution against a standard zinc solution in Chem 151. NaOH can be standardized by titrating its solution against a solution of a weighed amount of a primary standard acid. Several of these are used, but the most common is ***potassium hydrogen phthalate***, abbreviated KHP (even though there's no phosphorus in it!).

Make up a liter of NaOH solution at *about* 0.05 M concentration using the 50% aqueous NaOH on the side shelf and your 1-L plastic bottle. Note that the back cover of this manual has some useful information about concentrations and dilutions; you can use a graduated cylinder for the measurement of the 50% NaOH and the bottle itself is close enough to one liter. Cap both bottles tightly and invert the solution you've just prepared about ten times to mix it thoroughly. Keep it capped when you're not actually pouring from it, because it'll pick up CO_2 from the air and change concentration if it's open. And be sure to screw the cap back on the 50% NaOH when you're through with that bottle!

KHP is a nice dry crystalline solid that is a ***monoprotic*** acid—one replaceable proton per molecular formula. Its molecular weight (equal to its neutralization equivalent since it's monoprotic) is 204.23 g/mol. Figure out what weight of KHP would be equivalent to 40 mL of your 0.05 M OH^- solution, using the equations above. (It should be under a gram.) With this as a target weight, use the *top-loading balance* to weigh about five times that much into your weighing bottle. Use the *analytical balance* to weigh out four samples **by difference** (see page 12) into a 600-mL beaker and three clean, but not necessarily dry, Erlenmeyer flasks marked 1, 2, and 3 respectively. The samples can each be as much as 20% lighter than the target weight, but not more than about 10% heavier. What's important is to know accurately what the weight is to a tenth of a milligram, and to have a weight that can be titrated inside one 50-mL buretful.

Titrate the beaker sample quickly first to see about where the endpoint will be and what the endpoint looks like. Then do the three flask samples carefully, ***recording endpoint values to the nearest 0.02 mL***.

Dissolve each sample in about 50-100 mL of distilled water; exactly how much doesn't matter, because you're not titrating the water. Add about six drops of phenolphthalein indicator, and titrate rapidly at first, more slowly later when a pink spot is visible where the OH^- solution is being added from the buret. Stop at the first permanent pink color through the whole flask that doesn't fade after 30 seconds or so.

For your standardization titrations, *calculate a separate value of the molarity of the NaOH solution for each titration;* do the usual statistics, with a mean, standard deviation, standard deviation of the mean, and the relative standard deviation of the mean in ppt for the molarity. You should be able to get a standard deviation within about 3 ppt. Remember that you can use the Q test to reject suspicious values (see page 28). Do these calculations as soon as you've finished the titrations, so you can tell whether you need to run one or two more; the "3 ppt" target is important, because you won't be able to be sure about your unknown neutralization equivalent if you don't have a reliable molarity. Get a good molarity value before you go any farther!

For your unknown acid neutralization equivalent, weigh out just one sample by difference on the *analytical balance* again, but smaller— only about 0.1 to 0.15 grams. If your acid is water-soluble, just use water solvent, but if it requires ethanol use 1:1 water-ethanol mixture and warm it if necessary to get the sample dissolved.

The titration for your unknown acid won't use an indicator. Instead, you'll do a pH titration using a pH meter. You have a weak acid, and you're titrating with a strong base, so you can look up that kind of pH titration in Oxtoby (about page 340 in the 5th edition). The appearance of the titration curve (Oxtoby's Fig. 10.14) makes it clear that you have to take pH measurements out well past the endpoint. But if you have a *diprotic* acid, with two transferrable protons, there will be two endpoints (see Oxtoby's Fig. 10.17 on p. 352), but sometimes they won't be well resolved – how far out is far enough?

What you have to do is to take the titration out until the pH of the solution is 11 or close to it. That will make sure that you've titrated all the replaceable protons. Once you've dissolved your unknown acid sample, take it to a pH meter, but don't start the titration right away. The meter has to be calibrated by using buffer solutions with known pH. You'll use two *colored* standard buffers, which the meter will recognize. **Follow the standardization procedure printed on the meter.** Take a small clean beaker with you, to rinse the electrode into after taking it out of the buffer beaker; also take your squeeze bottle with a little deionized water for rinsing. *Always* rinse the electrode off with deionized water as soon as you take it out of any solution. And, of course, don't add any other solution to the buffer beakers.

Once the meter has been standardized, rinse the electrode off and place the electrode into the beaker with your unknown acid solution, and fill the buret with your 0.05 M NaOH solution. Fill it above the 0.00 mL line, and carefully drain it (into a small waste beaker) down to exactly 0.00 mL before beginning the titration. Make up a table in your notebook with two columns: *mL* OH^- and *pH.* You'll take quite a few readings, so leave lots of room. It may be a good idea to go to a new right-hand notebook page. Enter the first volume as *0.00* , and read the meter to enter the first pH value (probably somewhere around pH 3). Add 0.50 mL base, wait for the pH to stabilize (only a couple of seconds), and enter the second volume as *0.50* if that's the buret reading, and also enter the second pH value.

Keep adding approximate 0.5 mL increments from the buret and reading the pH after each addition. *Enter the buret volume to .02 mL each time,* along with the pH to .01 pH unit. Near the endpoint (or the first endpoint if you have a diprotic acid) the pH will change much more sharply, but don't worry about it; you can still use 0.5 mL increments. Keep adding base and taking pH readings until the pH reaches about 11.

Now use a spreadsheet program like Excel or Quattro Pro to plot your data: Enter your volume and pH data in two columns and plot pH against volume. Be sure the graph you get fills the paper and has useful ticks – every 0.2 pH unit and every 0.5 mL for volume. If you have a monoprotic acid, you should see one sharp break; the steepest part of the curve is the endpoint. On the other hand, if you have a diprotic acid you'll probably see two sharp breaks (though they might blend into an upward-sloping line before the break); the second should be at exactly twice the volume of the first. Use your ruler and tangents to the curve to establish where the endpoints should be. From the endpoint, of course, you can calculate a value for the neutralization equivalent for the acid. In general, if you have a small endpoint volume (15-20 mL) you probably have a monoprotic acid, while if you have a large endpoint volume like 35 or 40 mL you probably have a diprotic acid even if you didn't see an intermediate break in the pH curve (you won't see it if the pK_a's are close together).

How do you get the endpoint volume from your graph? It's the volume corresponding to the steepest rise in the pH curve, but how do you get that? Use tangents to establish where the point halfway up the rise is, and read that pH value (the curve is steep, so it's easy to read). Then use simple proportions to get the volume corresponding to that pH value: there's a tall skinny right triangle defined by your two data points *just before* and *just after* the pH break (see Irv Gratch's curve). Call those points (V_1, pH_1) and (V_2, pH_2). Now there's a smaller similar triangle defined by (V_1, pH_1) and the endpoint (V_{ep}, pH_{ep}), and V_{ep} is what you want. Write the proportion

$$\frac{V_{ep} - V_1}{pH_{ep} - pH_1} = \frac{V_2 - V_1}{pH_2 - pH_1}$$

Now just solve for V_{ep} and use that value to work out your NE value using the equation on page 132.

"But there's more!" as they say in TV ads.

Note in Oxtoby's Fig. 10.14 that halfway to the endpoint for a monoprotic acid, the acid is half neutralized so that the concentration of the acid HA is exactly equal to the concentration of its conjugate base A^-. This means that the numerator and denominator in the weak-acid K_a expression are the same, so that $[H_3O^+] = K_a$ or $pH = pK_a$. So you can read the pK_a for your acid right off the graph, halfway to the first endpoint.

Let's look at how Irv graphed his pH data and worked out his V_{ep} and pK_a. He used the spreadsheet to give him a BIG graph, one that filled the paper, didn't use data way past the endpoint (but far enough to get a clean upper tangent), and used only the part of the pH range that actually had data (for him, 2-12). The next page shows his plot, printed sideways to emphasize the need for a big graph.

First he draws in the lower, fairly flat tangent running along in his case from about pH 2.5 to 3.5. Then he draws in the upper tangent more or less parallel to the lower one – in his case, from about pH 10 to 11. Third, he draws in the most vertical tangent he can (but not a vertical line!) where the curve is most nearly vertical. This gives him two intersections between tangents, which on his graph are 126 mm apart (using a millimeter ruler). The endpoint is halfway between the two intersections, or 63 mm up from the bottom one. He's marked the endpoint with a big ×. Now he uses a horizontal ruler to find the endpoint pH *to the nearest 0.01 unit* (which is why you need ticks on the graph for every 0.02 pH unit). For him pH_{ep} turns out to be 6.72.

Now he uses the little proportion equation above:

$$\frac{V_{ep} - 17.00}{6.72 - 4.56} = \frac{17.50 - 17.00}{7.97 - 4.56}$$

or

$$V_{ep} = 17.00 + 0.317$$

$$V_{ep} = 17.32 \text{ mL}$$

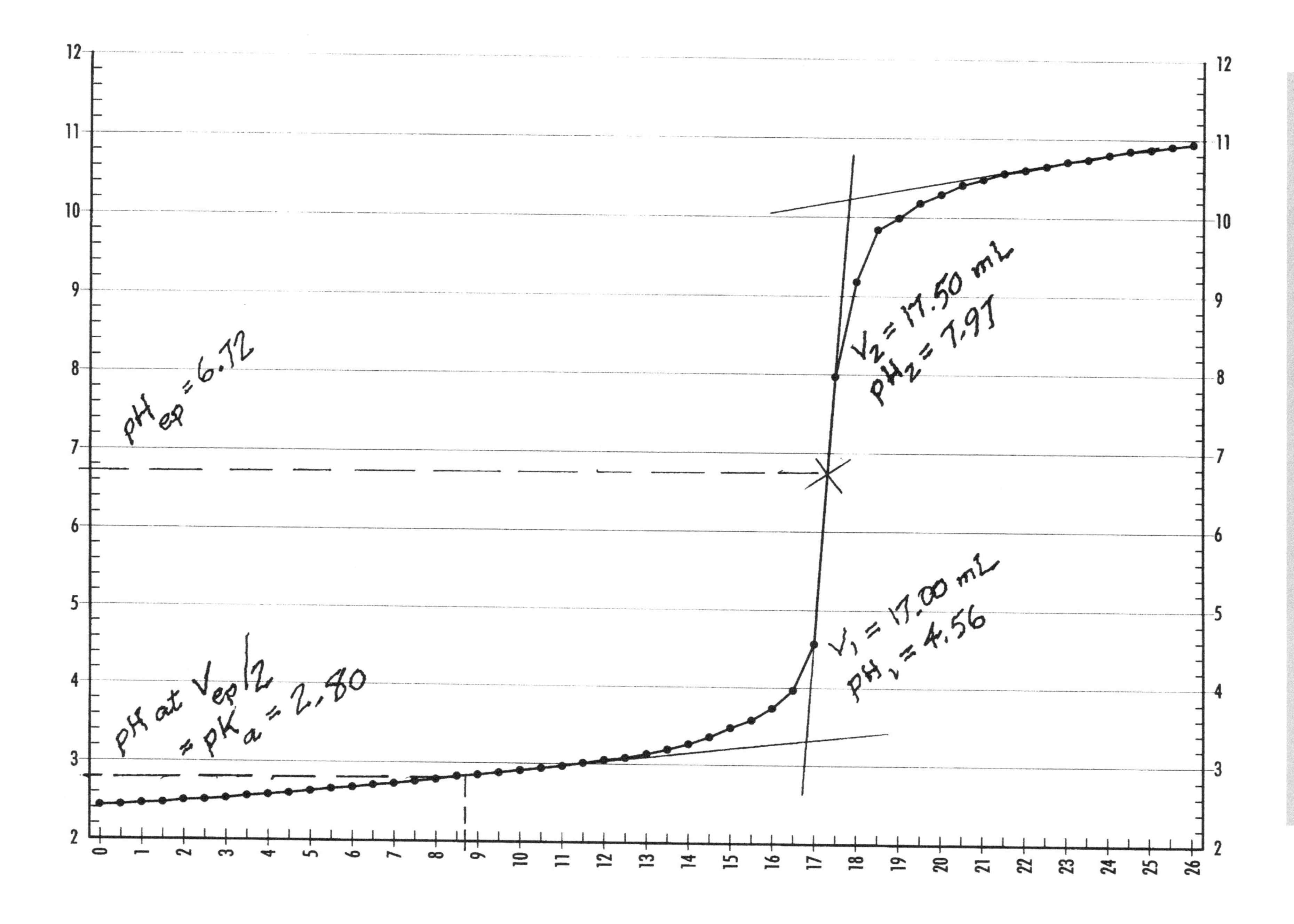

pH ep = 6.72
V2 = 17.50 mL
pH2 = 7.91
V1 = 17.00 mL
pH1 = 4.56
pH at Vep/2
= pKa = 2,80
2 3 4 5 6 7 8 9 10 11 12
0 1 2 3 4 5 6 7 8 9 10 11 12 13 14 15 16 17 18 19 20 21 22 23 24 25 26

Having a value for V_{ep} lets Irv calculate his neutralization equivalent ***and*** lets him calculate $V_{ep}/2$ – and knowing that lets him get his pK_a, in his case 2.80. So generate your NE and pK_a the same way.

This titration has given you two pieces of information about your acid, NE and pK_a, which you can compare against the unknown list. Sieve the list for the three quantities you have now: MP, NE, and pK_a . Even if this gets you down to a single acid, make up a little table with the *three* best matches, giving the acids' names, MP values, NE values, and pK_a values.

A bit more on titration curves: Irv had a monoprotic acid, but some unknowns are diprotic. For them, the titration curve has two possibilities. If pK_1 and pK_2 are well separated (maybe 4 units) you'll see two breaks, as in the left curve below, but if pK_1 and pK_2 are close together (only 1 or 2 units) there will only be one break because both COOH's are being titrated at the same time; the curve will slope up more sharply than for a monoprotic acid. This is shown in the right curve below – be alert!

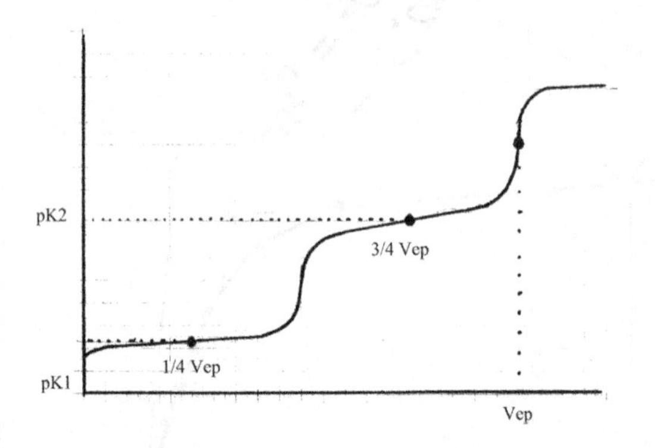

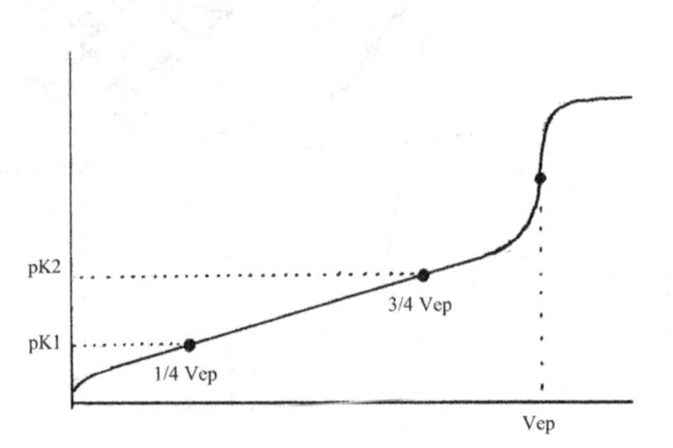

Notebook Writeup

Neutralization equivalent is another new experiment; use the usual headings in your notebook (but add the experiment title both in your notebook and in the table of contents):

- **Purpose (step 14, page 31)**
- **Procedure (step 15)**
- **Apparatus (step 16)**
- **Data (step 17 — tables of titration data)**
- **Calculated Results (step 19 — base molarity, NE of acid, pK_a value or values)**
- **Summary and Conclusions (step 20)**

^{1}H NMR FOR YOUR SOLID ACID (spectrum on notebook p. 74, interpretation on p. 75)

Once you have the three-acid table, bring your notebook with its short list to a faculty member and ask him for the ^{1}H NMR spectrum of your acid unknown. Use that spectrum to confirm your identification (or complete it). Be sure you write up your NMR interpretation in your notebook as you did for the liquid compounds, with a full interpretation! The proton NMR of the acid is a new experiment too, but this time there's no procedure; the "Calculated Results" will be your interpretation of the NMR spectrum. Mount the spectrum, folded once, on left-hand page 74 and do the interpretation on the opposite right-hand page75.

IR FOR YOUR SOLID ACID – ***optional, but read below*** (spectrum on nbk p. 76, interpretation p. 77)

There are some molecular-structure features that are hard to distinguish by NMR but easy by IR. In particular, you might have an acid that has a benzene ring and a Cl or a Br or a NO_2 group on it, and the neutralization equivalent isn't quite right for any of the three – and then in addition you're not sure whether the added group is in the 2-, 3-, or 4- position relative to the -COOH on the ring. IR is a big help here, and *if you want to* you can pick up the IR of your acid from a faculty member. If you get it you'll have to interpret it, but you can focus on Cl vs. Br vs. NO_2 or on *o*- vs. *m*- vs. *p*- substitution. It'll be another new experiment, and the "Calculated Results" will be your interpretation of the IR spectrum.

Acid	Formula (bold=ring)	N.E.	MP	pK_a
o-Anisic acid	$CH_3O\mathbf{C_6H_4}COOH$	152.15 g/mol	98-100°C	4.09
m-Anisic acid	$CH_3O\mathbf{C_6H_4}COOH$	152.15	106-108	4.08
p-Anisic acid	$CH_3O\mathbf{C_6H_4}COOH$	152.15	182-185	4.49
Azelaic acid	$HOOC(CH_2)_7COOH$	94.11	109-111	4.53, 5.40
Benzoic acid	$\mathbf{C_6H_5}COOH$	122.12	121-123	4.20
2-Bromo-2-methylpropionic acid	$BrC(CH_3)_2COOH$	167.01	44-47	3.99
2-Bromobenzoic acid	$Br\mathbf{C_6H_4}COOH$	201.02	148-150	2.85
3-Bromobenzoic acid	$Br\mathbf{C_6H_4}COOH$	201.02	155-158	3.81
4-Bromobenzoic acid	$Br\mathbf{C_6H_4}COOH$	201.02	252-254	3.99
2-Chlorobenzoic acid	$Cl\mathbf{C_6H_4}COOH$	156.57	138-140	2.88
3-Chlorobenzoic acid	$Cl\mathbf{C_6H_4}COOH$	156.57	155-157	3.83
4-Chlorobenzoic acid	$Cl\mathbf{C_6H_4}COOH$	156.57	239-241	3.99
trans-Cinnamic acid	$\mathbf{C_6H_5}CH{=}CHCOOH$	148.16	133-134	4.44
2,4-Dimethoxybenzoic acid	$(CH_3O)_2\mathbf{C_6H_3}COOH$	182.18	108-110	3.44
2,5-Dimethylbenzoic acid	$(CH_3)_2\mathbf{C_6H_3}COOH$	150.18	132-134	3.99
3,5-Dinitrobenzoic acid	$(O_2N)_2\mathbf{C_6H_3}COOH$	212.12	205-207	2.85
Glutaric acid	$HOOC(CH_2)_3COOH$	66.06	95-98	3.77, 6.08
Glycine	H_2NCH_2COOH	37.53	240 *(dec)*	2.35, 9.78
3-Hydroxybenzoic acid	$HO\mathbf{C_6H_4}COOH$	138.12	201-203	4.08
4-Hydroxybenzoic acid	$HO\mathbf{C_6H_4}COOH$	138.12	215-217	4.58
Itaconic acid	$HOOCCH_2C({=}CH_2)COOH$	65.05	166-167	3.85, 5.45
Maleic acid	$HOOCCH{=}CHCOOH$	58.03	140-142	1.91, 6.33
DL-Malic acid	$HOOCCH_2CH(OH)COOH$	67.03	131-133	3.46, 5.10
Malonic acid	$HOOCCH_2COOH$	52.03	135-137	2.83, 5.70
Mandelic acid	$\mathbf{C_6H_5}CH(OH)COOH$	152.15	119-121	3.41
Nicotinic acid	$\mathbf{NC_5H_4}COOH$	123.11	236-239	4.75
2-Nitrobenzoic acid	$O_2N\mathbf{C_6H_4}COOH$	167.12	146-148	2.18
3-Nitrobenzoic acid	$O_2N\mathbf{C_6H_4}COOH$	167.12	140-142	3.46
4-Nitrobenzoic acid	$O_2N\mathbf{C_6H_4}COOH$	167.12	239-245	3.44
Oxalic acid	$HOOCCOOH$	45.02	190 *(dec)*	1.27, 4.27
Phenylacetic acid	$\mathbf{C_6H_5}CH_2COOH$	136.15	77-79	4.31
o-Phthalic acid	$HOOC\mathbf{C_6H_4}COOH$	83.07	205 *(dec)*	2.95, 5.41
Pimelic acid	$HOOC(CH_2)_5COOH$	80.08	103-105	4.48, 5.42
o-Salicylic acid	$HO\mathbf{C_6H_4}COOH$	138.12	158-161	2.98
Succinic acid	$HOOCCH_2CH_2COOH$	59.04	188-190	4.21, 5.64
L-Tartaric acid	$HOOCCH(OH)CH(OH)COOH$	75.04	169-170	2.99, 4.34
o-Toluic acid	$CH_3\mathbf{C_6H_4}COOH$	136.15	103-105	3.90
m-Toluic acid	$CH_3\mathbf{C_6H_4}COOH$	136.15	108-110	4.27
p-Toluic acid	$CH_3\mathbf{C_6H_4}COOH$	136.15	180-182	4.36

SUMMARY FOR SOLID ACID

By the time you finish this project, you've done several experiments on your solid acid and written each experiment up separately. But there needs to be a final summary that gives an all-inclusive view of what you've done and identifies the acid once and for all. Take a new right-hand page in your notebook and head it "***Unknown Acid Summary.***" Give a brief summary (half a page to a page) of the sequence of experiments you went through and how they gradually led you to an identification of the acid.

FINAL SUMMARY WRITEUP

You should already have summarized in your notebook the experimental data that support your identification of the two organic liquids, and separately but in much the same form the solid acid. But there needs to be a summary of all the work you've done for the semester. Go back to the short overall summaries you've written for the organic liquids and for the solid acid, and extract data from them to give a one-page final summary as the end of your notebook.

Use a separate table of data (BP and key spectral data) for each liquid, and another table (MP, NE, and NMR data) for the solid acid. For each of the three compounds you choose, sketch the molecular structure and write the name as it appears in the appropriate table.

At the very end of the writeup, copy the pledge in the form indicated on page 8 and sign it. Of course we expect you to mean it, so if there are any reasons for you to be concerned about the pledge (like having written stuff on loose pieces of paper instead of the notebook) explain them.

The next page shows Irv Gratch's final summary, which is better than most of his work.

87

I.V. Gratch
4/12/65

SPRING SEMESTER PROJECT — Final Summary

(53) A mixture of 2 organic liquids was separated by fractional distillation (GC purity: LB 98%, HB 97%)

Low-BP component:	^{13}C NMR	IR	^{1}H NMR
BP 80-81°C	15 ppm 37 44 216	2978 cm^{-1} 1719 1370 1169	a 0.89 ppm / triplet / area 1.59 b 2.03 singlet 1.57 c 2.47 quartet 1.00
2-butanone BP 80°C	4 pks 216 ppm C=O ketone	2978 open chain C-H 1719 ketone C=O	singlet isolated CH_3 triplet/quartet $-CH_2CH_3$ areas 3:3:2

High-BP component:			
BP 170-173°C	16, 19, 29, 121, 124, 130, 140 ppm	3020 cm^{-1} 2870 815	a 1.45 / doublet / area 2.28 b 2.51 / singlet / 1.12 c 3.07 / septet / 0.34 d 7.31 / doublet / 1.50
p-cymene	7 pks, 4 in 120-140 benzene ring region	3020 arom C-H 2870 chain C-H 815 p-subst. benzene ring	septet/doublet isopropyl d aromatic areas 6:3:1:4

(96) Solid Acid	NE	pK_a	^{1}H NMR
MP 182-184°C	153.6 g/mol	4.61	a 3.81 ppm / singlet / area 1.69 b 6.92 / doublet / 1.00 c 7.85 / doublet / 1.02
p-anisic acid MP 182-185°C	152.15 g/mol	4.49	a fits CH_3-O- b+c fit p-disubst benzene

CHECKOUT (Must be complete by the end of last working lab period–that's *before* notebook is due!)

It's very important that the desks used in lab this semester be ready to use next semester—so important that we won't grade your notebook unless you've checked out! ***FOLLOW THESE INSTRUCTIONS EXACTLY, AND IT'LL BE QUICK:***

1) Take everything out of your desk and check each item against the desk-contents sheet on the next page. Separate the stuff into three groups:

   Group I: All the returnable items on the desk-contents sheet. Check out anything that's missing, cracked, or broken so you have a complete set in good shape.

   Group II: Any extra equipment or glassware that you've checked out from the stockroom during the semester. Return this to the stockroom and get it crossed off your apparatus checkout sheet in the stockroom file.

   Group III: All the nonreturnable, non-reusable stuff like matches or dirty corks **AND** leftover chemicals. Throw these away, remembering that chemicals must be disposed of in the ***HAZARDOUS WASTE*** containers except for ethanol, which can go down the sink; return usable stuff to the appropriate place.

2) Clean all the stuff in Group I *thoroughly*; wash and dry all the glassware that's not already spotless. Shake out the towel in the bottom of your drawer and spread it out in the drawer again.

3) When you're sure you've completed steps 1 and 2, get your original apparatus sheet from the stockroom and ask a lab assistant or the stockroom manager to check you out.

4) Return the apparatus to your desk, and lock the desk.

5) The stockroom manager will give you two chits when you've finished checking out. The first will authorize the business office to return the appropriate amount of your breakage deposit; keep it for them. The second should be taped inside the back cover of your lab notebook before you turn it in to be graded. **We won't grade any notebook that doesn't have the checkout chit in it!**

Made in the USA
Monee, IL
05 April 2025

15183908R00079